AF661409

A TEXTBOOK ON

FRUITS AND PLANTATION CROPS

A TEXTBOOK ON

FRUITS AND PLANTATION CROPS

By

V. Ponnuswami, Ph.D.
Former Dean (Horticulture)
Professor and Head (Horticulture)
Vanavarayar Institute of Agriculture,
(Affiliated to Tamil Nadu Agricultural University)
Manakkadavu, POLLACHI-642103 (T.N.)
e-Mail: swamyvp200259@gmail.com

M. Kumar, Ph.D.
Assistant Professor (Horticulture)
Vanavarayar Institute of Agriculture,
(Affiliated to Tamil Nadu Agricultural University)
Manakkadavu, POLLACHI-642103 (T.N.)
e-Mail: kunshorts@gmail.com

V. Arumugam, Ph.D.
Assistant Professor (Horticulture)
Vanavarayar Institute of Agriculture,
(Affiliated to Tamil Nadu Agricultural University)
Manakkadavu, POLLACHI-642103 (T.N.)
e-Mail: vrarumugam@gmail.com

S. Ramesh Kumar, Ph.D.
Assistant Professor (Horticulture)
Vanavarayar Institute of Agriculture,
(Affiliated to Tamil Nadu Agricultural University)
Manakkadavu, POLLACHI-642103 (T.N.)
e-Mail: maaduphd@gmail.com

C. Krishnamoorthy, Ph.D.
Assistant Professor (Horticulture)
Vanavarayar Institute of Agriculture,
(Affiliated to Tamil Nadu Agricultural University)
Manakkadavu, POLLACHI-642103 (T.N.)
e-Mail: plantdoctorkrishna@gmail.com

JAYA Publishing House

DELHI-110095 (INDIA)

Copyright © 2015, Jaya Publishing House, Delhi (India)

All rights reserved. Neither this book nor any part may be reproduced or used in any form or by any means, electronic or mechanical, including photocopying, microfilming, recording, or information storage and retrieval system, without the written permission of the publisher and author.

The information contained in this book has been obtained from authentic and reliable resources, but the authors/publisher cannot assume responsibility for the validity of all materials or the consequences of their use. The authors/publisher have attempted to trace and acknowledge the materials reproduced in this publication and apologize if permission and acknowledgements to publish in this form have not been given. If any material has not been acknowledged please write and let us know so that we may rectify it.

First Published in 2015

Reprint in 2017

ISBN: 978-93-84337-26-1

Published by :

JAYA PUBLISHING HOUSE

Publisher and Distributor

27-B, Pocket-B, Dilshad Garden

Delhi-110095 (INDIA)

Ph.: 91-11-22597616, 91-9891277233

Email: info@jphindia.com
jayapublishinghouse@gmail.com

Printed in India

Laser Typeset by Amrit Graphics, Shahdara, Delhi-110032

Contents

Preface

THE Horticulture sector encompasses a wide range of crops namely fruit crops and plantation crops. Horticulture has emerged as an indispensable part of agriculture, offering a wide range of choices to the farmers for crop diversification. It also provides ample opportunities for sustaining large number of agro-industries which generate substantial employment opportunities. The horticulture sector contributes around 28% of the GDP from about 13.08% of the area and 37 % of the total exports of agricultural commodities. India as the second largest producer of fruits and vegetables, largest producer and consumer of cashew nut, tea spices, third largest producer of coconut, forth largest producer and consumer of rubber and sixth largest producer of coffee in the world. Today, as a result synergy between focused research, technological and policy initiatives, high efficiency inputs, horticulture in India, has become a sustainable and viable venture for the small and marginal farmers. Besides, the sector has also started attracting entrepreneurs for taking up horticulture as a commercial venture. Therefore, there is a great scope for the horticulture industry to grow and flourish. The growth of any enterprise on a national scale pre-supposes sound and responsive infrastructure. During the past several years, the country has created infrastructural facilities to the requirements of horticultural research, education and development in the country in terms of setting up of Institutes, establishment of separate Departments of Fruits, Vegetables, Floriculture in several State Agricultural Universities and carving out State Departments of Horticulture from the erstwhile Agriculture Departments in many of the States. India occupies first place in production of mango, banana, litchi, papaya, pomegranate, sapota and aonla; 2nd place in limes and lemons and 5th place in pineapple production of the world. The productivity of grape is highest in the world. India has higher national average productivity in banana and sapota compared to world average productivity, while in citrus, mango, apple, guava, pineapple, papaya, India has substantially low productivity in comparison to the world average. Plantations Crops, There are two groups of plantation crops namely coconut, arecanut, oil palm, cocoa, cashewnut dealt by Ministry of Agriculture and tea, coffee, rubber dealt by the Ministry of Commerce. The combined volume of output of tea, coffee and rubber is about 2 million MT and constitutes about 3 per cent share of the total value of agricultural output in India.

In this book covers as production technology of tropical and plantation crops grown in the world have been discussed. In this book covers thirty two leading chapters' *viz.,* scope and importance, global and national scenario, nutritive values classifications, planting systems, good agricultural practices and production technology of fruits and plantation crops. Although these are discussed in separate chapters, they maintain a fair continuity. The various aspects have been illucidated in a very lucid language for the benefit of students and other professional bodies. Photographs, charts and tables have been used wherever necessary, for clarity and to provide due weightage to the topics in the context of world acreage, production as well as importance.

This book has been written with the hope that it may provide a satisfactory guidance would be of much use undergraduate students of Agriculture and Horticulture, extension workers, orchard development workers and also farmers of the country would be benefit greatly from the information contained in this book.

I also express appreciation to all colleagues and co-workers who have assisted me in any way in preparation of the manuscript of this book. In particularly I am thankful to all my colleagues in the department of horticulture for inspiration and making valuable suggestions for improvement in the text. I would like to express my deep appreciation to Jaya Publishing House, New Delhi for handling the manuscript expeditiously and bringing out the book in a nice get-up within a short span. The authors are thankful to Tamil Nadu Agricultural University and Vanavarayar Institute of Agriculture.

Since this is a first edition, it may contain some errors that crept in inadvertently in spite of all editorial care. I am always interested to hear comments and suggestions from readers for improvement. You can contact me by e-mail, for which address is: kumshorts@gmail.com

Authors

CHAPTER

1

SCOPE AND IMPORTANCE OF FRUITS

SCOPE OF FRUIT GROWING

Fruit cultivation contributes to the health, happiness and prosperity of the people. The standard of living of the people can be judged by production and consumption of fruits per capita.

Fruit has been a major food for mankind from time immemorial. Fruits and vegetables are also called as 'protective foods' because of their unique nutritional and medicinal properties. Even the Primitive man, the food habit was based mainly on fruits and also tubers which were found wild in the forests. In fact Adam who is considered to be the progenitor of man was said to have temptingly given the apple to Eve, the first woman.

In Ramanaya and Mahabaratha the cultivation and consumption of fruits were described. In the ancient Tamil literature the importance of fruits and that of three significant nutritious fruits called 'Mukkanigal' maa (mango), pala (jack fruit) and vazhai (banana) have been described. The importance of amla the 'nellikani' in bestowing longevity was brought out by a reference to kinh 'Adhiyamaan' giving it to the Tamil poetess 'Avvayar' who was a great scholar and should live long.

I. Importance of Fruit Growing

In India, fruit crops are cultivated in an area of 6.98 million ha with on annual production of 44.04 million tonnes. India contributes to 12.6 per cent of the total world production. The requirement of fruits has been calculated as 40-100 gram per head per day.

Cultivation of fruit crops plays an important role in the prosperity of any nation. It is generally stated that the standard of living of the people can be judged by per capita production and consumption of fruits. Let us see the advantages of growing fruit crops in comparison with cereals, pulses and oil seeds.

Fruit crops are capable of giving higher tonnage per unit area than other field crops. For example Wheat crop produces on an average of 12 – 15 tonnes/ha (two crops/year) and Hybrid Rice can give only a maximum of 25 tonnes/ha (three crops/year). While the fruit crops, Banana can yield 35-40 tonnes/ha, Papaya 100 - 150 tonnes/ha in 2½ years (40-60 tonnes/year), Mango 25 tonnes/ha, grapes 60 – 80 tonnes/ha in two harvests per year under tropical climate. Apple can give 60 – 80 tonnes/ha in a medium density planting system, while pine apple can give as high as 100-130 tonnes/ha in a crop duration of 1½ years under high density planting (HDP) system.

If we consider that the heavy tonnage may not be the true measure to compare the crops in question, the average calorific value they produce per unit area can be taken as a better measure for the above purpose. When two crops of wheat produces 52 lakhs calorie from an area of one hectare in a year, one crop of banana generates 370 lakhs calorie from the same area and the calorie of papaya would be 450 lakhs calorie per hectare per year.

For the water requirement point of view except banana all the fruit crops are requires less water than the field crops like rice. If we follow fertigation in fruit crops, using lesser quantity of water and fertilizers we can also produce higher amount of fruit production.

The potential of fruit crops in the growth of national economy is noteworthy. The prosperity of the country lies in building up its foreign exchange reserves. Being a country having varied climatic conditions ranging from tropical to subtropical and to temperate, India has very immense potential for the production of different fruits and their export.

The following points are the importance of fruit growing in nutshell:

1. Fruit growing provides better foods as they are good source of vitamins and minerals and have medicinal properties.
2. Fruit crops give high energy return per unit area compared to other crops as discussed above.
3. Fruit growing is remunerative proposition
4. Fruits are basic need for many kind of industries/factories like canning, preservation, dehydration, essential oils etc.
5. Fruit culture encourages horticultural based industries like fruit processing and preserved product like jam, jelly, squash and canned fruits.
6. Fruit growing helps in the development of package, transport, refrigeration, oil and wine industries. Mulberry plantation is associated with sericulture and apiculture is dependent of fruit crops.

7. Fruit growing creates jobs to rural poor.
8. Waste lands can be utilized by growing hardy fruit crops (arid zone fruit crops) like ber, amla or aonla, bael, dates, custard apple etc.
9. Air pollution can easily minimized by massive planting of fruit species because trees are know to be the best air purifier.
10. Fruit growing (tree fruits) can also check the global warming. Now a day global warming is the series problem arising. On the other hand the rice fields produce more amount of methane gas which causes for global warming.\

SOCIAL IMPORTANCE OF FRUIT CROPS

Offer Nutritional Security

- Fruit requirement @ 100 g/day/adult, vegetables @ 300 g/day/adult
- Protective foods, vit 'A', 'B', 'C', Iron and Phosphorous
- Vitamin 'A' being derived from fruits & veg.– 60%, of the total requirement
- Vitamins 'C' derived from fruits & veg. – 90% of the total requirement
- An adult requires 11 lakh calories/year
- 0.44 ha of wheat, 0.03 ha of banana and 0.16 ha of Mango – will provide 11 lakh calories
- Mineral value: Ca, P, Mg, K, Fe.

Health: Take an Apple a day; keep the doctor away and Laxative value.

Measure of Quality of Life

Consumption of fruits is a measure of standard of living.

Symbol of Prosperity

- All function with flowers and fruits
- Celebrations with flowers and fruits

Environment friendly – Growing of fruit trees

High Productivity

- Productivity of fruit plants per unit area is higher than the agricultural crops.
- Production of grapes is 50 tons, pineapple 45 tons, papaya 40 tons, banana 25 tons per hec.
- High productivity of fruit trees gives higher economic return per unit area.

Highly remunerative from horticultural crops

Grapes	–	1.5 lakh / ha
Citrus	–	1 lakh/ha
Banana	–	80,000/ha

Suited to Dry and Wastelands

- Tamarind, Amla, Annona etc., are drought tolerant
- Salt tolerant trees are Sapota, Guava etc.,

Offer Scope for Value Addition

Post harvest – Jam, Jelly, Juice, etc.,

High potential to earn dollars – e.g. Crop like Cashewnut

High contribution to GDP i.e. 29.5% (from 8.5% area under horticultural crops).

Dietary uses – eg. Papaya rich all vitamins and minerals

Others

- Less labour requirement
- Less input and less care

Multiple Cropping Systems

- Mango, litchi and pear are the permanent crops and come into bearing late. The filler trees like kinnow, peach, plum, papaya, phalsa can be intercropped in the permanent trees during prebearing period. Vegetables can be intercropped between the tree rows in the orchard.
- Multiple cropping system in orchards generate more avenues for employment apart from giving additional income.

Higher Profit

- Fruit crops give much more production per unit of area and are certainly more paying than ordinary farm crops.
- Growers can net an income of Rs.30000 - Rs.60000 per acre by growing fruit crops.

Trading

- Fruit production has a great possibility in trade and industrial development.
- Processing units, garden tools, CFB cartoons and wooden crates, baskets, spray pump, insecticides and pesticides, fungicides, cold storage, refrigerated vans are the important areas of trade.

Diversification

Fruit crops fit well in diversification programme. Wheat, rice rotation is not profitable. Water table is going deep. Rice crop should be replaced slowly with fruit crops. Fruit crops can be grown successfully on waste-land, backyard of home, tube wells etc.

❑❑❑

CHAPTER

2

GLOBAL AND NATIONAL SCENARIO OF FRUITS

GLOBAL SCENARIO

Globally, China ranks first in area, production. India and Brazil ranks 2nd and 3rd respectively. The details are given in the following Table

S. No.	Country	Area (million hectares)	Production (Million MT)	Productivity (MT/ha)
1.	China	11.83	137.066750	11.6
2.	India	6.98	81.285334	11.6
3.	Brazil	2.33	38.368678	16.5
4.	United States of America	1.14	26.548859	23.3
5.	Indonesia	0.80	17.744411	22.3
6.	Philippines	1.24	16.370976	13.2
7.	Mexico	1.26	15.917806	12.7
8.	Turkey	1.11	14.974561	13.6
9.	Spain	1.54	13.996447	9.1
10.	Italy	1.13	13.889219	12.3
11.	Others	27.93	270.594597	9.7
	World	**57.27**	**646.757638**	**13**

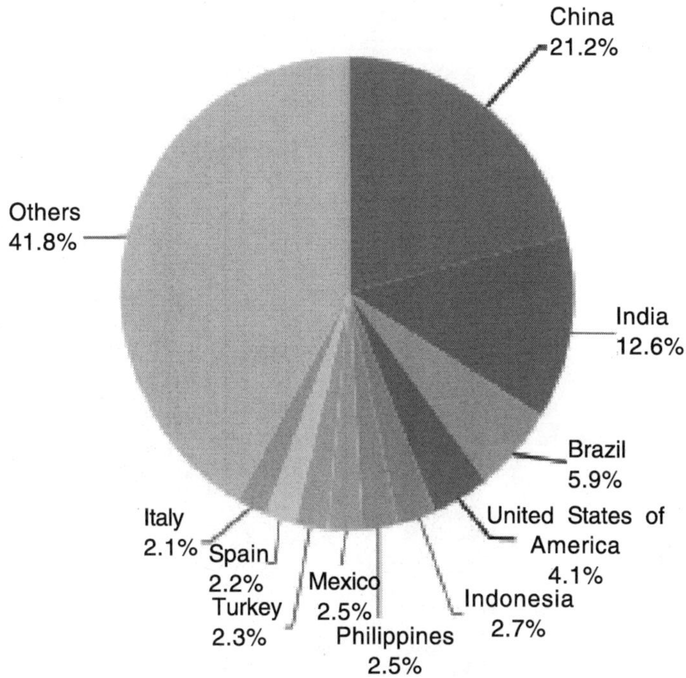

Fig. A. Global scenario of fruit crop percentage

(**Source:** FAO, 2014 and Indian Horticulture Database, 2013)

National scenario

Year	Fruits		Plantation crops	
	Area ('000 ha)	Production ('000 MT)	Area ('000 ha)	Production ('000 MT)
1991-92	2874	28632	2298	7498
2001-02	4010	43001	2984	9697
2002-03	3788	45203	2984	9697
2003-04	4661	45942	3102	13161
2004-05	5049	50867	3147	9835
2005-06	5324	55356	3283	11263
2006-07	5554	59563	3207	12007
2007-08	5857	65587	3190	11300
2008-09	6101	68466	3217	11336
2009-10	6329	71516	3265	11928
2010-11	6383	74878	3306	12007
2011-12	6705	76424	3577	16359
2012-13	6982	81285	3641	16985

Area and production of Fruit crops in India (2012-13)

S. No.	Crop	Area ('000 ha)	Production ('000 MT)	Productivity (MT/ha)
1.	Banana	776	26509	34.2
2.	Mango	2500	18002	7.2
3.	Citrus	1042	10090	9.7
4.	Papaya	132	5382	40.7
5.	Guava	236	3198	13.6
6.	Apple	312	1915	6.1
7.	Pineapple	105	1571	14.9
8.	Sapota	164	1495	9.1
9.	Grapes	118	2483	21.1
10.	Pomegranate	113	745	6.6
11.	Litchi	83	580	7.0
12.	Others	1402	9315	6.6

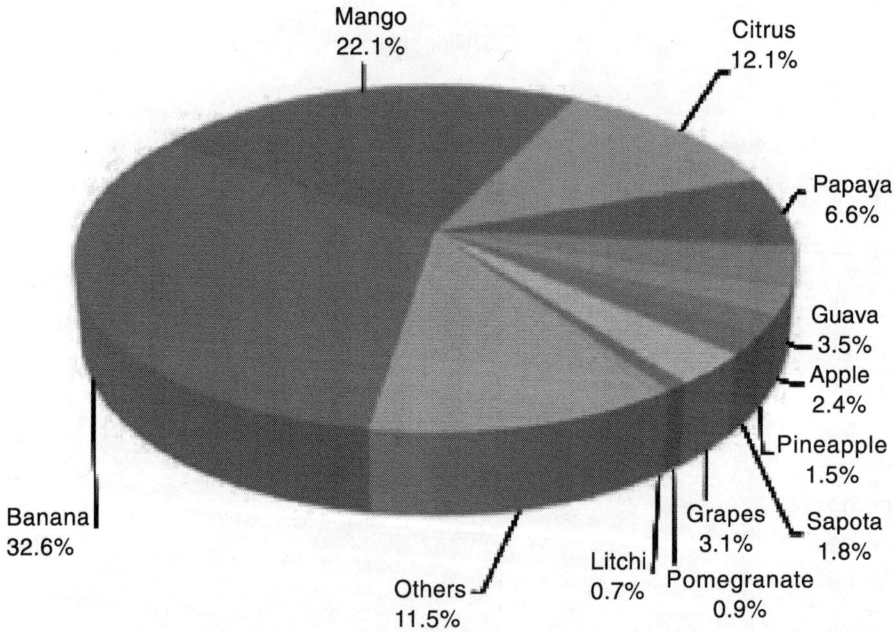

Fig. B. Production share of major fruit crops in India (2012-13)

(**Source:** FAO, 2014 and Indian Horticulture Database, 2013)

❑❑❑

CHAPTER

3

NUTRITIVE VALUE OF FRUITS

Fruits are the chief source of minerals, vitamins, carbohydrates, fats and protein. These constituents are essential for the normal physiological wellbeing and help in maintaining healthy state through development of resistance against pathogens. As per dietic requirements, one should consume at-least 90-100g fruits per day in addition to vegetables, pulses, meat, eggs, milk etc. to maintain proper health.

NUTRITIVE VALUE OF FRUITS IN HUMAN HEALTH

The aspect which we cannot afford to neglect is the protective role of fruits in our diet. Fruits are the rich source of vitamins and minerals.

I. Vitamins

Vitamin A

Fruits are rich source of vitamins as mentioned above. For example mango, papaya and jack fruit have the important constituent the beta carotene which is actually the precursor of vitamin A. Its deficiency will lead to xerophthalmia, the drying of cornea. In extreme cases, a softening of cornea known as keratomalacia is noticed. In this instance, normal vision gets affected. Vitamin A deficiency will leads to night blindness. Mango and papaya fruits have very good source of readily available beta-carotene, 1990 μg per 100 g in mango and 880 μg per 100 g in papaya.

Vitamin B complex

Vitamin B_1

Vitamin B complex another important group of vitamins. Vitamin B_1 thiamine is an important member of B group. Prolonged deficiency of thiamine leads to beri

-beri disease. Banana, pine apple, plum apricot and almond can supplement vitamin b1. since fruits are consumed without cooking, this vitamin is readily available.

Vitamin B_2

The next important member of B complex is the vitamin B_2 or riboflavin. As a part of coenzyme, it is essential for several oxidation processes inside the cell and is mainly concerned with energy and protein metabolism. Important symptoms due to inadequate intake of this vitamin in the diet are glossitis (soreness of tongue), angular stomatitis (cracking at the angles of the mouth) and respiratory infections. Fruits like jack, papaya, custard apple, bael passion fruit and banana can help significantly in supplementing the vitamin to our diet.

Vitamin C

Vitamin C or ascorbic acid is an essential vitamin. It is a strong reducing agent and is involved in the collagen synthesis and bone and teeth calcification. Vitamin C deficiency causes scurvy characterized by weakness, bleeding gum and defective bone growth. Vitamin C helps in absorption of dietary iron by keeping it in the reduced ferrous form (Fe^{++}). Fruits are rich source of vitamin C, oranges and pineapple contain 35-40mg of this vitamin per 100g of pulp while papaya contains 57 mg per 100g of pulp and ripe guava contains 212 mg per 100g of pulp.

Amla supplies as high as 600 mg of vitamin C for every 100 g of pulp. This fruits unique importance is vitamin C is doesn't degrade during cooking. West Indian cherry having higher amount of vitamin C. It is estimated that 1400 mg per 100g of pulp.

PHYTOCHEMICALS AND ANTIOXIDANTS

According to recent research results many Phytochemicals found in fruits act as a powerful antioxidants protecting cells and organs from damage caused by free radicals, neutralizing their damaging effects. They are the biologically active substances in plants that give them colour, flavour, odour and protection against not only diseases affecting the plants but also human being.

Oranges, lemons, lime and grape fruits besides being principal sources of vitamin C and phytochemicals called limonoids. This antioxidant has been found to be very effective against cancer. Blueberry is the number one antioxidant fruit. Grapes show anti-inflammatory, anti-thrombic and antioxidant effects.

Minerals

Dietary need of important minerals like iron, calcium and phosphorus also is very well taken care of by the consumption of fruits. Black currants, a group of temperate fruits are the riches source of iron (8.5 mg / 100g of pulp) followed by dried dates 7.3 mg, custard apple 4.3 mg pineapple – 2.4 mg /100g of pulp. Iron is an important constituent of blood protein, the hemoglobin (in red blood corpuscles) which is involved in the process of cell respiration.

Calcium and phosphorus are the two minerals which go hand in hand in the formation and maintenance of the skeletal system. Pomegranate – 70mg, wood apple – 110 mg, avocado – 80 mg in 100 g of edible pulp. Temperate fruits like raspberry, black currants contain very high phosphorus (110 mg for every 100g of pulp) banana – 36 mg dried apricot and date 110 - 120 mg of calcium per 100g of edible portion. Wood apple 130 mg, lime 90 mg and fig 80 mg of calcium.

Medicinal Value

Fruits have certain medicinal value. Generally fruits are found to possess laxative effect. A ripe banana, papaya fruits and figs are having a good laxative property. The latex of the skin of unripe papaya fruit contains a proteolytic enzyme papain which is used in meat tenderization and preparation of certain digestive medicine. A mixture of lime juice and ginger also cures pittham characterized by vomiting and giddiness.

Sweet orange will reduce very high fever in the body. Indian goose berry or amla or aonla is used for its curative property. It is used for minor disease like cough, bronchitis, dyspepsia and complicated disease like jaundice and diabetes, combined extract of amla and bitter gourd keeps a check on the blood sugar level of diabetic patients.

A recent scientific report from France revealed that jack fruit could be very useful in the treatment of the most dreaded diseases AIDS (Acquired Immune Deficiency Syndrome). A chemical called 'jacaline' present in the jack extract could protect the healthy lymphocytes from infection by AIDS virus.

❑❑❑

CHAPTER

4

CLASSIFICATION OF FRUITS

The fruit crops are classified according to their habitat as woody perennials like trees, shrubs or vines and herbaceous plants. The woody perennials may be either deciduous or evergreen. The deciduous plants have a distinct rest period, which coincides with winter. But the evergreen plants do not have any distinct rest period, retaining leaves and be inactive during that period.

1. CLASSIFICATION BASED ON HABITAT

- Woody perennials (trees, shrubs or vines)
  - Evergreen (No distinct rest period in winter and remains active)
  - Deciduous (Distinct rest period coincides with winter and remains inactive)
- Herbaceous perennials

1. FRUITS BORNE ON WOODY PLANTS

The woody plants may be trees, shrubs or vines. The woody plants may be either evergreen or deciduous.

(i) Tree fruits

(1) Evergreen (Tropical and Sub-tropical)

a) Tropical: Mango, Sapota and Guava.

b) Sub-tropical: Mangosteen, Litchi and Acid lime.

(2) Deciduous (Temperate)

a) Pome (Fleshy fruit): Apple, pears and Quince.

b) Drupe (Stone fruits): Plum, Peach and Apricot.

(ii) Small fruits

 (1) Deciduous: Raspberry and Blackberry.

 (2) Evergreen: West Indian cherry.

2. FRUITS BORNE ON HERBACEOUS PERENNIAL PLANTS

Fruits borne on herbaceous perennial plants may be further classified as those with prostrate growth and upright growth.

(i) Prostrate growth: Strawberry.

(ii) Upright growth: Banana and Pineapple.

3. Classification of Fruits Based on Temperature

Based on the temperature requirements, the fruit crops are classified as tropical, subtropical and temperate fruits. Further in tropical types, they are divided as semiarid and arid fruits.

CLASSIFICATION BASED ON TEMPERATURE

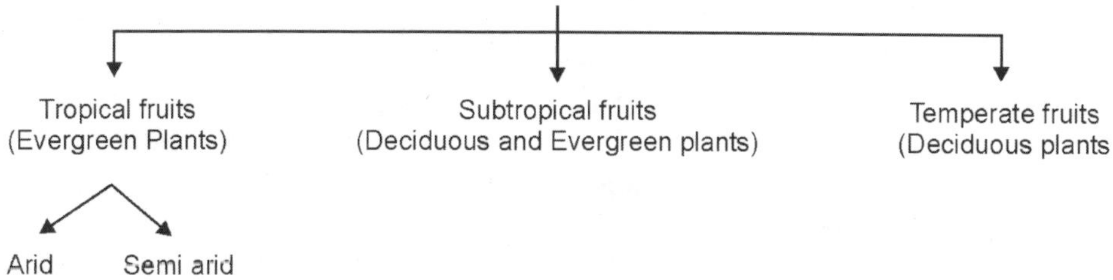

4. CLASSIFICATION OF FRUITS BASED ON THE TOLERANT TO THE RELATIVE HUMIDITY (RH) OF THE ATMOSPHERE

Based on the tolerant to the relative humidity of the atmosphere, the fruit crops can also be classified as arid, semi-arid and humid zone fruit crops.

Arid Zone Climate

The arid / semi-arid regions have peculiar eco-climatological features and they can exist in tropics, sub-tropics as well as temperate zones also. In these zones, a deficiency of moisture restricts but not necessarily inhibits plant growth. The rainfall in this zone is very low (0-250 mm) and is confined to 2 to 4 ½ months and the remaining are dry months. Besides, in these regions, higher solar radiation incidence (450-500 cal/cm^2/day) and high wind velocity (20 km/hr) results in a high potential evapotranspiration (6 mm / day). The soils are of poor quality, with low fertility level and poor water holding capacity (20 to 25%) and with high infiltration rate (9cm). Ber, annona, date palm, phalsa are typical examples of arid or semi-arid fruit crops.

Arid Zone Fruits

Aonla, ber, pomegranate, date palm, custard apple, jamun, bael, wood apple and manila tamarind.

Semi-Arid Zone

A Semi-arid climate or steppe climate describes climatic regions that receive low annual precipitation, usually (but not always) between 250 - 500 mm of precipitation. This region exhibits low and erratic rainfall, low humidity and high temperature. Fruits of arid region can be cultivated in this zone also Mango, Sapota, Guava, Jack, Avocado, Ber, Pomegranate and Tamarind etc.

Humid Zone

This zone is distinct from sub-tropical zone in the sense that temperature during winter will not go down very low, similarly it is distinct from arid or Semi-Arid zone as the atmosphere humidity will be always higher. This kind of peculiar climate is prevailing in South India in western and eastern ghat foot hills ranging from 300 to 1200 m M.S.L (Mean Sea Level) and also in the valleys and foot hills in North Eastern hill regions.

Humid Zone Fruits

These fruits are generally grown in places receiving more rainfall (>1200 annually) and having higher relative humidity (RH). Pineapple, jack, Mangosteen, avocado, litchi and mandarin oranges are examples for humid zones fruits.

5. CLASSIFICATION BASED ON RATE OF RESPIRATION

Climacteric Fruits

In this group the respiration rate will be increased after harvesting. So, after harvesting the fruits will be easily ripened. More technically, in climacteric fruit 'ripening' is controlled by the fruits production of Ethylene and a significant increase in CO_2 production. E.g. Mango, Banana, Papaya, Guava, Sapota, Fig, Annona, Apple, Pear, Plum, Peach, Apricot, Kiwi, Passion fruit etc.,

Non Climacteric Fruits

In this group the respiration rate will be arrested after harvesting. So, there is no ripening process after the harvest of the fruit and allow the fruits to ripen on tree (or) attached with the plant. Non climacteric fruit produce little or no Ethylene and no large increase in CO_2 production. In order to improve external skin colour and market acceptance, citrus like orange, lemon, mousambi and kinnow can be treated with ethylene, as a de-greening agent. Ethylene treatment breaks down the green chlorophyll pigment in the exterior part of the peel and allows the yellow or orange carotenoid pigments to be expressed.

E.g. All citrus fruits (Acid lime, Lemon, Mandarin orange, Sweet orange, Grapefruit, etc.,), Grapes, Pomegranate, Lichi, Watermelon, Cherry, Raspberry, Blackberry, Strawberry etc.

6. CLASSIFICATION BASED ON BEARING HABIT

On the basis of bearing habit, fruit trees are classified in to eight categories to facilitate cultural operations like training, pruning etc.

- Fruit buds bore terminally and giving rise to inflorescence without leaves. e.g. Mango, Cherry, etc.
- Fruit buds borne terminally and unfolding to produce leafy shoots which terminate in flower clusters. e.g. Apple, Pear, etc.
- Fruit buds borne terminally and unfolding to produce leafy shoots with flowers or flower clusters in the leaf axils. e.g Guava.
- Fruit buds borne laterally containing flower parts only and giving rise to inflorescence without leaves or leaves present, they are reduced in size., e.g. Citrus, Coconut, Papaya, Coffee, etc.
- Fruit bud borne laterally and unfolding to produce leafy shoots terminally in flower clusters this type of flowering is noticed in grapes and cashewnut.
- Fruit buds borne laterally unfolding to produce leafy shoots with flower clusters in the leaf axils. E.g. Fig, Avacado, etc.
- Fruit buds borne both terminally and laterally but unfolding to produce inflorescence terminally. E.g. Walnut.
- Fruit buds always borne adventitiously in old trunk or shoots or cauliflorous bearing (Etymology - From Latin *caulis, stem + flôs, flôr-, flower.* The production of flowers and fruits on the older branches or trunks of woody plants). E.g. Jack fruit, Cocoa, Indian star gooseberry.

7. CLASSIFICATION BASED ON FRUIT MORPHOLOGY

Simple fruit

Berry: Banana, Papaya, Grape, Sapota and Avocado.

Modified Berry

Sl. No.	Type of Fruit	Examples
1	Balusta	Pomegranate
2	Amphisarca	Wood apple, Bael
3	Pome	Apple, Pear, Loquat, Quince
4	Hesperidium	Citrus fruits
5	Nut	Litchi, Walnut, Rambutan
6	Capsule	Aonla, Carambola

Drupe (Stone): Mango, Peach, Plum, Ber, etc.

Aggregate fruits: Eteario of berries - Custard apple and Raspberry.

Multiple fruit: 6.4.1. Syconus - Fig.

6.4.2. Sorosis - Jackfruit, Pineapple, Mulberry, Bread fruit etc..

8. CLASSIFICATION BASED ON PHOTOPERIODIC RESPONSES

Short day plants	Long day plants	Day neutral plant
Strawberry, Pineapple, Coffee, etc.	Passionfruit, Banana, Apple,etc.	Papaya, Guava, etc.

9. CLASSIFICATION BASED ON RELATIVE SALT TOLERANCE

Highly sensitive	Medium Tolerant	Highly tolerant
Mango, Apple, Citrus, Pear, Straw berry, etc.	Pomegranate, Cashew, Fig,Jamun, Phalsa, etc.	Datepalm, Ber, Amla,Guava, etc.

❑❑❑

CHAPTER

5

PLANTING SYSTEM AND SPECIAL OPERATIONS ADOPTED IN FRUIT CROPS

System of Planting

- It should aim at providing maximum number of trees per hectare
- Adequate space for proper development of the trees and ensuring convenience of cultural operations.
- The choice of system of planting depends upon the topography, crop, variety, plant density, production technology to be followed thereafter.

Square System

- In this system a tree is planted at each corner of a square.
- Whatever may the Planting distance.
- The distance between row to row and plant to plant is same.
- Most commonly followed and simplest of all and easy to lay out.
- Irrigation channels and paths can be made straight.
- The possibility of cultural operations in two directions is the greatest advantage of this system
- Better supervision of the orchard is possible as one gets a view of the orchard from one end to the other.

Rectangular system

- Trees are planted on each corner of the rectangle.
- Trees planted in straight rows running at right angles.

- Distance from plant to plant and row to row not the same.
- This system is very easy to lay out in the garden.
- The wider alley spaces available between rows of trees permit easy intercultural operation and even the use of mechanical operations.

Diagonal/ Quincunx system

- This system of planting is similar to square with the addition of a tree in the centre of each square as filler of intercrop or the same crop.
- It accommodates about 1.5 to 2 times or (89%) double the number of plants in square system but does not provide equal spacing.
- The filler tree chosen may be a short lived one.

Hexagonal System

- Triangles are equilateral.
- Trees are planted at corners of equilateral triangles.
- These way six trees from a hexagon with the seventh tree in the centre.
- Distance from tree to tree in six directions remains the same.
- It accommodates 15% more trees than square system.

Triangular System

- The trees are planted as in square system but the difference being that those in the even numbered row or mid-way between than those in the odd rows instead of opposite to them.
- Difficult to lay out and not of much practical importance in traditional plantations.
- It is used in high density planting in Amarapali mango to accommodate 1600 plants per hectare.

Contour System

- Used in hills side plantings, trees are adjusted as per available contour.
- When slope exceeds 10%, terraces are made to plant the tree.
- The main purpose of this system is to minimize land erosion and to conserve soil moisture so as to make this slope fit for growing fruits and plantation crops.

Hedge Row Planting

- It is generally followed in high density planting of apple and pineapple.
- It is followed for mechanized fruit cultivation.

Planting Density

- Spacing also determines the intercultural operations to be followed in the orchard.
- The mechanization of the intercultural operations is also based on the distance between and within the rows.
- The distance between trees and rows varies with the system of planting, fruit crop and varieties to be planted.
- The vigour of the plant determines the spacing.

Planting Distance and Density of Various Fruit Crops

Spacing (m)	Tree number / ha	Fruit crops planted
2 x 2	2500	Dwarf papaya and dwarf banana
3 x 3	1111	Tall banana, tall papaya and grape
5 x 5	400	Peach, fig and pomegranate
7 x 7	204	Peach, plum and guava
9 x 9	121	Sapota
10 x 10	100	Mango, cherry and date palm
12 x 12	68	Jamun and tamarind

- Proper layout of the orchard would facilitate easy supervision, management and planning for future expansion.
- Cultivation of perennial and annual crops of fruits, vegetables and to some extent flower should also be taken into account while making layout.

The different categories of planting densities and management systems in use today are:

1. Low density planting: Trees are widely spaced, number ranging between 100 and 250 trees/ha. This is a non-intensive system of growing fruits.
2. Medium- density planting: Tree spacing is reduced accommodating 250-500 trees/ha.
3. High density planting: Trees are spaced very close together. The number of trees may range from 500 to 10000/ha.

High Density Planting

- High density planting refers to growing of crops in closer spacing thereby to accommodate more population per unit area of land.
- Planting of fruit trees rather at a closer spacing than the recommended one using certain special techniques with the sole objective of obtaining maximum productivity per unit area without sacrificing quality is often referred as HDP. Eg. Banana, Mango.
- High-density planting helps to increase the yield/unit area. Besides higher yield, HDP also helps to reduce labour cost and increase the efficiency of input utilization. E.g. Mango (Amrapali) and Jack (PLR1).

Types of High Density Planting in Apple

- Low HDP (< 250 trees /ha)
- Moderate HDP (250-500 trees/ha)
- High HDP (500-1250 trees/ha)
- Ultra high HDP (> 1250 trees/ha)
- Recently, super HDP system has been also established in apple orchards with a plant population of 20,000 trees/ha.
- In some orchards, still closer planting of apple trees is followed (70,000 trees/ha), which is often referred as meadow orchards.

SPECIAL OPERATIONS IN FRUIT CROPS

Ringing

Removal of a complete ring of bark from a branch or trunk to increase fruit bud formation e.g., grapes

Notching

Partial ringing above a dormant lateral bud –removal of a small narrow strip of bark just above and close to a dormant bud. eg grapes, apple

Bending

Practiced in guava, where large branches are bent to induce dormant buds.

Smudging

Practice of smoking the trees like mango to produce off-season crop

Deblossming

Mango plants may produce flowers even after one year of planting. The inflorescences are removed upto three years.

Desuckering

In the initial stage of growth in banana, side suckers are removed at monthly interval.

Tipping

Removal of floral remnants a week after opening of the last hand in banana.

Propping

Propping is providing support with bamboo or casuarina poles to avoid damaged by wind. Banana plants at flowering are propped.

Use of Bunch Covers (skirting)

Banana bunch is covered with polythene sleeves to avoid sunscald and to improve the quality of fruits.

Mattocking

Removal of pseudostem after harvesting the banana bunches for favouring the growth of the follower sucker.

Thinning

- Removal of a part of flower buds, flowers or fruits with the objective of increasing the marketability of the remaining fruit and reducing the alternate bearing tendency.
- Excessive bearing in grape vines can be reduced by cluster thinning or berry thinning.

Top working

- Old orchards are renovated by top working i.e., by grafting with popular scion materials.
- Renovation of old country pear orchards through top working with quality varieties is done in pear orchards.

Use of Growth Regulators

- Dipping the inflorescence at calyptra fall stage at 25 ppm GA_3 and again at pepper stage increases the size of berries.
- Spraying of NAA at 20 ppm at flowering stage reduces flower drop. Eg. Thompson seedless variety / other seedless varieties.

Soil Solarisation

- Soil temperature can be increased by 5-10°C by covering the field with thin transparent plastic sheets.
- It checks long wave black radiation from soil and prevents loss of energy by hindering moisture on evaporation.

Canopy Management

- Canopy management is the manipulation of tree canopies to optimize the production of quality fruits.
- Components like tree training and pruning, affects the quality of sunlight intercepted by trees, as tree shape determines the presentation of leaf area to incoming radiation.
- An ideal training strategy centers around the arrangement of plant parts, especially, to develop better plant architecture that optimize the utilization of sunlight and promotes productivity.

Principles of Canopy Management

- Maximum utilization of light.
- Avoidance of built- up microclimate congenial for diseases and pest infestation.
- Convenience in carrying out the cultural practices.
- Maximizing productivity with quality fruit production.
- Economy in obtaining the required canopy architecture.

❑❑❑

CHAPTER 6

GOOD AGRICULTURAL PRACTICES OF FRUITS

There is a need to develop fully-integrated orchard management systems that will promote production and be environmentally sound. Healthy and vigorous orchards produce high-quality fruit at the best possible cost and also, reduce the need for chemical treatments.

Best management practices for orchards include attention to: site preparation, soil management, water management including irrigation and drainage, nutrient management and pest management. Growers can adjust each component to maximize profits while protecting the environment.

Orchard Site Preparation

When planning a new orchard, select and prepare an appropriate site at least one to three years in advance. Consider soil testing, past levels of nematodes, organic matter levels, perennial weed control, drainage, soil depth, slope, stoniness and frost pockets. Soil testing is a must prior to planting. Determine nutrient and pH levels and correct any problems. Control nematodes, especially Root Lesion nematode. This is crucial to proper establishment of young fruit trees. Nematodes can damage roots and allow fungi to enter roots, disrupting water and nutrient absorption. To determine whether fumigation is necessary, look at the previous crop (corn, for example, increases nematodes), soil type (sandy soils tend to have higher populations than clays), rootstocks tolerance to nematodes and the results of soil samples. If counts are higher than 1,000 nematodes per kilogram of soil, treatment is recommended. Plan ahead – consider soil test results, past levels of nematodes, weed control, drainage, soil depth, slope, stoniness and frost pockets.

Fumigation

Applying fumigants is usually done with a three-point hitch cultivator which places fumigants in a shallow band 1.75 m wide and 15 cm deep. The entire field can be fumigated or just the strips where trees will be planted. Before applying fumigants, prepare a good seedbed. A new method uses a twin-shank sub soiler to deliver fumigant in a narrow band at 15, 30 and 45-centimetre depths. Establishing the sod cover in the summer before fumigation is recommended. Fumigating row strips through the sod allows better weed and erosion control. This may give better nematode control and also sub soils the planting area. The reduced tillage also preserves organic matter and reduces erosion.

Tree Density

Deciding how wide the tree rows should be and how far apart trees should be planted will affect productivity, nutrient management, pest management and water requirements. Before making a decision, consider equipment requirements, availability of skilled labour and availability of irrigation water.

Apples

Tree density has steadily increased over the years as dwarfing rootstocks replace standard rootstocks. The most cost-efficient systems in use are high-density training systems, such as slender spindle (1,750 trees per hectare) are in use. The advantages are:

- Earlier production with higher yields
- Orchard efficiency is higher (more fruiting wood is produced per hectare)
- Production costs per bin decrease
- Potentially higher-quality fruit
- Pesticide use may decline (tree row volume techniques)
- Cost recovery time is shorter

This System Requires

- High initial investment
- More professional skills and management are needed

Peaches

The standard for Ontario is 417 trees per hectare. This allows easy movement of standard equipment. The slender spindle system allows densities of 834 trees per hectare. Research completed in 1991 shows yields up by 17% compared to the standard system. Consider the following when making a decision:

- Higher costs to establish orchard
- Pruning methods will be different
- Training of trees is critical in the first and second years 75% of all work can be done from the ground

Soil Management

Good soil management in orchards should promote tree growth and good health, productivity and overall fruit quality while preserving soil structure. Issues include ground covers, organic matter and erosion. Soil management systems include clean cultivation, cultivation plus cover crop, sod plus herbicide strip, sod plus mulch and intercropping between tree rows. In Ontario, growers usually use sod or cultivation plus cover crop. Clean cultivation decreases organic matter, degrades soil structure, increases erosion and increases the potential for winter injury.

Cultivation/Cover Crops

Soil is worked in April and cultivated regularly until early June. Cultivation reduces competition for moisture between trees, grasses and weeds and increases the air in the soil and soil temperatures (which may help reduce risk of spring frost). In mid-June, a cover crop is planted. When cultivating an orchard, leave some plant material on the soil. The purpose of cultivation is to suppress annual weed growth, not to overwork the soil.

Factors to consider when deciding on the cover crop include:

- Ease of establishment
- Dry matter produced
- Effect on nematodes and pests
- Nutrient interactions

The cover crop most widely used is annual ryegrass. It establishes quickly and will survive droughts by delaying establishment until conditions improve.

Sod Systems

Producers grow permanent sod between tree rows and mow sod for the life of the orchard. Advantages are:

- Decreased erosion
- Moderate soil temperatures
- Increased organic matter

- Decreased mechanical injury of roots
- Water penetrates soil more easily
- Easier orchard operations
- Decreased soil compaction

Some growers are trying to establish sod the year before planting. In the fall, sod in the tree row is killed with an herbicide. The following spring, trees are planted into the dead grass without cultivation.

Herbicide Strips

The objective of herbicide strip is weed suppression during the critical growth stage from early spring to midsummer. A strip of bare ground is left at the base of the trees to reduce competition for moisture between trees and grasses and to aid in the control of voles and mice. The wider the strip, the better tree growth will be. However, a permanently bare strip creates soil problems, increases the possibility of roots being injured over the winter and encourages perennial weeds. The best solution is to use mulches. Mulches are organic materials that are placed within the tree row. Mulches should be applied early to allow decomposition before fall months.

Advantages

- Moisture is retained /conserved.
- Soil temperatures are moderated.
- Microbial activity is higher.
- More extensive rooting is encouraged.
- Soil structure improves.
- Enhanced nutrient availability.

Disadvantages

- Mulches may encourage rodents.
- Material and labour increases costs.
- Potential for excessive nitrogen.
- Introduction of weed seeds.
- Mechanical harvest of fall apples more difficult.

Possibilities for mulch include: straw, hay (legume hay may contain high levels of nitrogen which may increase late tree growth causing winter damage), wood chips and related products, decomposed organic wastes and grass clippings. Apply mulches when soil moisture is high, usually in the spring.

Soil Compaction

The constant movement of equipment between tree rows may compact soil and result in poor drainage. Sub-soiling or mechanical aerators open up soil. However, care must be taken to prevent unwanted root pruning. Techniques should be used when soils are dry as working on wet soil will make problems worse. Some growers are modifying their mowers to throw sod clippings into the row area as mulch. A strip of bare ground at the base of the tree helps to reduce competition for moisture from the sod and aids in vole and mice control.

❑❑❑

CHAPTER 7

MANGO

Botanical name: *Mangifera indica* L.,

Family: Anacardiaceae

Chromosome no.: 2n = 40

Origin: Indo - Burma region (India and South East Asia)

Mango is one of the most important among the tropical fruit crop of India and world and having socio-economic significance. Mango undoubtedly deserver to the "National fruit of India".It is the leading fruit crop of India, very popular and considered to be the choicest of all fruits grown in India. It is the favorite fruit of all parts of India and has been repeatedly acclaimed as "King of Fruits", "King Tropical of Fruits" and "Ambassador of fruits", owing to delicious quality of fruit (Because of its rich, luscious, aromatic flavor and a delicious taste in which sweetness and acidity are delightfully blended),richness in vitamins and minerals and liking to Indian palate. The ripe fruit is fattening, diuretic and laxative. It is also called as 'bathroom fruit' because of it laxative property. Kernel is effective against diarrhoea and asthma. In Tamil Nadu, the mango is considered, along with banana and jackfruit, as one of the three royal fruits (Mukkani: Ma-pala-vazhai) occupying first place in terms of sweetness and flavor.

Nutritive Value

Mangoes are full packed with vitamins, minerals and anti-oxidants and contain like all fruits very few proteins, fats and calories. Mango fruit is rich in pre-biotic dietary fiber, vitamins, minerals, and *poly-phenolic flavonoid* antioxidant compounds. *polyphenolic anti-oxidant* compounds in mango are known to offer protection against breast and colon cancers. Mango fruit is an excellent source of Vitamin-A and flavonoids like *beta-carotene, alpha-carotene,* and *beta-cryptoxanthin.* 100 g of fresh fruit provides 4800 IU recommended daily levels of vitamin-A.

Climate and Soil

Mango can grow from sea level to an altitude of about 1500 MSL, However it cannot be grown on a commercial scale in areas above 600 m. The temperature between 24 and 27°C is ideal for its cultivation. High temperature during fruit development and maturity gives better quality fruits. Rainfall range between 25cm and 250cm. For normal flowering, mango requires dry but cool winter during flowering. Dry cool winter during flowering followed by hot summer and good rainfall after harvest is highly ideal for better mango cultivation. Low temperatures (13°C – 19°C) are good for flower bud differentiation. After flowering rainfall is require for fruit development. Moderate humidity coupled with water stress or rest for 2-3 months before flowering improves flower bud formation. High rainfall prior to flowering induces excessive vegetative growth and arrest the flowering. Temperature, also governs the flowering and ripening of fruits.

Soil

The mango can be grown on a wide range of soils except clay, extremely sand, rocky calcareous, alkaline and water logged. It cannot tolerate saline conditions. It is a large tree with deep tap-root system. It can be grown from alluvial to lateritic soils except in black cotton soil having poor drainage. Mango prefers slightly acidic soils with a pH range of 5.5 to 7.0. Soils having good drainage are ideal for mango.

Varieties

In India, about 20 - 30 cultivars are being grown commercially. However, only a few varieties are commercially cultivated throughout India. There are 2 types of mango varieties.1. Monoembryonic 2. Polyembryonic varieties

1. **Monoembryonic varieties:** Presence of one embryo in a seed is called as monoembryony. Mono-embryonic mangoes consist of a single embryo which gives rise to a single hybrid plant not true to the type (i.e. do not have the characteristics of parents) therefore, they are raised by asexual propagation. Almost all the varieties of India are mono-embryonic.
2. **Polyembryonic varieties:** Presence of more than one embryo in a seed is called as polyembryony. The phenomenon of polyembryony is known to occur in a number of mango varieties. Seedlings arising from the adventitious embryo of nucellar origin are highly uniform. Poly-embryonic varieties are Bappakai, Bellary, Chandrakaran, Dwarf, Goa, Kurukkan, Olour, Kasargod, Mazagoan, Mylepalium, Nakari, Nileshwar dwarf, Pahutan, Peach, Prior, Salem, Starch and Vellaikulumban.

Alphonso (Syn: Gundu, Khader, Badami, Hafus)

One of the most popular variety of India and exported to other countries. It is mainly grown in Ratnagiri area of Maharashtra and to a small extent in parts of south Gujarat and Karnataka. Pulp is firm, fibreless with excellent orange colour. It has good sugar:acid blend. Keeping quality is good. It is susceptible to spongy tissue.

Banganapalli (Syn: Baneshan, rajapalayam sappattai, Safeda)

A widely cultivated,early-maturing mango of south India. It is the main commercial variety of Andhra Pradesh. Its fruits are large-sized. The pulp is fibreless, firm and yellow with sweet taste. Fruits have good keeping quality.

Dashehari

One of the most popular variety of north India, it is a mid-season mango. Fruits are medium-sized, with pleasant flavour, sweet, firm, and fibreless pulp. Stone is thin and keeping quality good.

Langra

An important commercial mango variety of north India, it is biennial-bearer and a mid-season variety, with good quality fruits. Flesh is firm, lemon-yellow in colour and scarcely fibrous. It has characteristic turpentine flavour. Keeping quality is medium.

Neelum

A heavy-yielding, late-season mango in south India, it has regular-bearing habit. Fruits are medium-sized with good flavour. Flesh is soft, yellow and fibreless. Keeping quality is good.

Bangalora (Syn: Kallamai, Kilimookku, Totapuri)

Widely grown in south India, it is a regular and heavy-bearing mango. Fruits are medium to large with prominent sinus. Fruit quality is medium. Flesh is cadmium-yellow and fibreless.

Paiyur 1

This variety is a selection from Neelum trees grown in Karukkanchavadi of Dharmapuri District developed at Regional Research Station, Paiyur of Tamil

Nadu Agricultural University, Tamil Nadu. Trees are dwarf in stature and suited for high density planting with 400 plants per hectare.

Bombay Green (Syn: Malda)

It is commonly grown in north India due to its early fruiting quality. It is also called Malda in Utter Pradesh and northern India and Sehroli in Delhi. It is a pollinating variety and also has highest content of vitamin C. Fruits have strong and pleasant flavour. Pulp is soft and sweet. Fruit quality is good and keeping quality is medium. It is a biennial bearer and highly susceptible to malformation. It is a very early season variety.

Chausa (Syn: Khajri, Samar Bahisht)

It is a late variety maturing towards the end of July or beginning of August. It is one of the sweetest variety of mangoes lacking acidity. Biennial bearing, susceptibility to mango malformation and very vigorous growth habit are its major drawbacks.

Fazli

A mango variety originated in Bihar and West Bengal. Flavour is pleasant and pulp is sweet and fibre less. Keeping quality is good. The tree is very vigorous and medium to heavy cropper with biennial bearing habit. It has moderately good keeping quality.

Gulabkhas

It is the foremost commercial variety of Bihar and has spread recently in West Bengal and U.P. It is favoured for its characteristic rose flavour and very sweet taste. The flesh is firm and fibreless. This is a heavy yielder but biennial bearing. Both fruit quality as well as keeping quality is very good.

Himayuddin (Syn. Himam pasand or Imampasand)

It is a choice variety and an irregular bearer. Fruits are large in size with oblique shoulders. Flesh is firm, fibreless and lemon chrome in colour. Fruits have good taste, flavor and keeping quality.

Himsagar

This variety is most popular and indigenous to Bengal. This is one of the choicest varieties of Bengal and has gained extensive popularity. Fruit is sweet in taste; quality and keeping quality are good. It is an early season variety.

Jehangir

It is also a choice variety. Fruits are large with high quality and ovate in shape. it is a shy and irregular bearer.

Kalepad (Karu Neelum)

It is also a commercial variety. Fruits resemble Neelum in all respects except in the skin colour which is dark green turning to light green after complete ripening.

Kesar

This is a leading variety of Gujarat with a red blush on the shoulders. It has good processing quality. Keeping quality is good. It is an early season variety.

Mulgoa

This is a commercial variety of south India. It is quite popular among the lovers of mango owing to high quality of its fruit. It is a late season variety with a biennial bearing habit. This variety is mono embryonic in India and poly embryonic in Florida. It is a late season variety with a biennial bearing habit.

Nadusalai (Syn: Peter, Peter-Prasad, Raspuri, Piari, Grape and Yerra Goa)

It is one of the commercial varieties of Western India. The bearing is heavy and fruit quality is good. It is a heavy bearer with biennial bearing habit. It is an early variety maturing around mid-May and June.

Rumani

Fruits are medium round and apple shaped. This variety has lowest number of perfect flowers (0.74 %). It has good keeping quality. It is a heavy yielder. The fruit quality is moderately good. The fruits have a good keeping quality.

Sendhuram (Syn: Swarnarekha, Chinna swarnarekha, Sindhuri, Sundari (in the Delhi market) and Lal Sundari)

Fruits have attractive pinkish red blush on the shoulders. Fruit quality is medium and keeping quality is good. It is an early season variety and a biennial bearer. It is a very early variety found in Andhra Pradesh. The bearing is heavy but biennial in habit.

Vanraj

It is a highly prized variety of Vadodra district of Gujarat and fetches good returns. Fruit quality and keeping quality are good. It is a mid season variety.

Niranjan

A clonal selection developed at Marathwada Agricultural University, Parbhani, Maharashtra and released in the year 1984. A regular and off-season bearing variety, it flowers during June-July and matures in October.

Mango hybrids and their characters

Hybrid	Parentage	Important characters
Indian Agricultural Research Institute, New Delhi		
Mallika	Neelum x Dashehari	Regular-bearers, high TSS, good colour, uniform fruits, moderate keeping quality
Amrapali	Dashehari x Neelum	Dwarf, regular-bearers, cluster-bearing, small-sized fruits, good keeping quality and suitable for high density planting
Regional Fruit Research Station, Vengurla		
Ratna	Neelum x Alphonso	Semi-dwarf, regular in bearing, good keeping quality. Fruits are free from spongy tissue.
Sindhu	Back cross progeny of Ratna (Ratna x Alphonso)	Stone is extremely thin(seedless mango), regular in bearing, free from spongy tissue & fibreless
Indian Institute of Horticultural Research, Bangalore		
Arka Aruna	Banganapalli x Alphonso	Regular bearer, dwarf, HDP (9400 plants/ ha),400g-attractive reddish tinge-skin, Pulp-fibreless, TSS 22°brix,Dwarf, regular, prolific & medium bearer. Free from fibre and spongy tissue.
Arka Puneet	Alphonso x Banganapalli	Regular-bearer, attractive skin colour, medium-sized, free from spongy tissue. Good keeping quality, good sugar, acid blend
Arka Anmol	Alphonso x Janardhan Pasand	Regular yellow colour, medium-sized (200-300 g), free from spongy tissue. Good keeping quality, suitable for export.
Arka Neelkiran	Alphonso x Neelum	Semi vigorous (7 x7m)-204, HDP, Medium size-260g, redblush, pulp deep yellow, spongy tissue free-export, keeping quality good

[Table Contd.

Contd. Table]

Hybrid	Parentage	Important characters
Fruit Research Station, Anantharajupet-Kadapa District, Andhra Pradesh		
Neeleshan	Neelum x Baneshan	Late maturing and Good quality.
Neelgoa	Neelum x Yerra Mulgoa	Regular bearing, high yielding' good keeping quality and late maturing
Swarna Jahangir	Chinna suvarnarekha x Jahangir	Medium yielder, fruits are attractive in colour, good transporting quality and late maturing
AU Rumani	Mulgoa x Rumani	Heavy and regular bearer with good keeping quality & late season
Fruit Research Satation, Sangareddy, Medak District, Andhra Pradesh		
Manjeera	Rumani x Neelum	Precocious, regular bearer and heavy yielder
Horticultural College and Research Institute, Periyakulam, Tamil Nadu		
PKM 1	Chinna swarnarekha x Neelum	Mid-season, Regular, cluster, long distance transport, 250-300g, prominent beak
PKM 2	Neelum x Mulgoa	Mid-season, 650-700g, keeping quality good, glands prominent
Agricultural Research Station , TNAU, Paiyur, Tamil Nadu		
Paiyur-1	Open pollinated Selection from Neelum	Dwarf, HDP,400 plants/ha,150-200g
Agricultural Research Institute, Sabour, Uttar Pradesh		
Prabhasankar	Bombay x kalaepad	Semi dwarf, regular in bearer, fruit is of good keeping quality. Strong tendency to bear fruits in adverse weather conditions.
Mahamudbahar		Bombay x kalaepad Tree is semi dwarf and regular in bearing.
Jawahar	Gulabkhas x Mahamooda Bahar	Semi dwarf, highly regular in bearing &precocious
Central Research Institute of Sub-tropical Horticulture, Lucknow, Uttar Pradesh		
Ambika	Amrapalli x Janardhan Pasand	Regular, high yielder, yellow colour with red blush on peel, late variety
Arunika	Amrapali x Vanraj	Dwarf in stature; can be planted at 5m x 5m (400 plants per hectare) Regular bearer; tolerant to anthracnose
MPKV, Rahuri, Maharashtra		
Sai- Suhandh	-	Improvement-Totapuri, free from spongy tissue and stone weevil, export variety, table, flesh firm

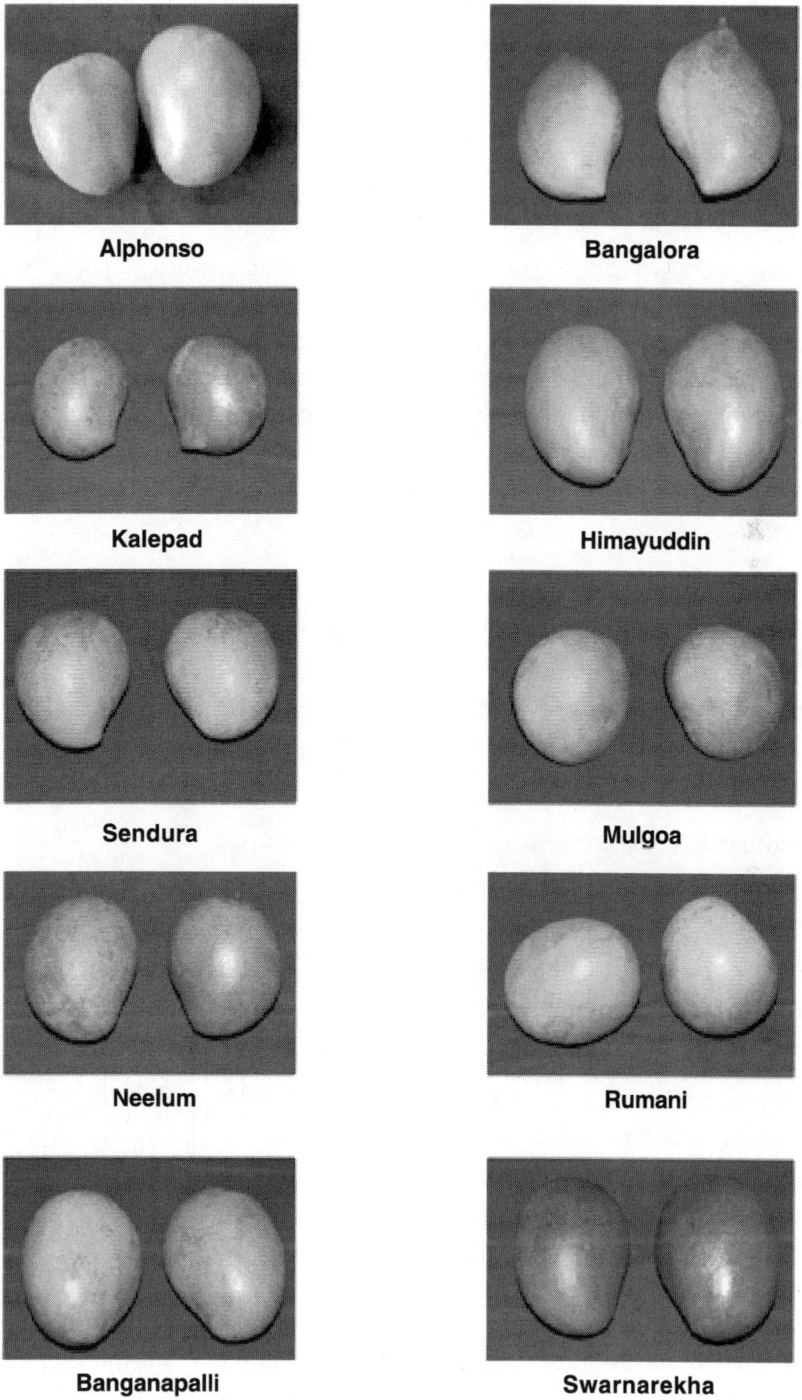

Fig. 1. Different mango varieties

Propagation

Mango is a highly heterozygous and cross pollinated crop. There are two types *i.e.* Monoembryonic and Polyembryonic. Polyembryonic varieties can be propagated through seed because they produce true to type seedlings whereas the monoembryonic varieties need to be propagated vegetatively.

Inarching or approach grafting is the principal method of propagation followed by veneer grafting. Cuttings do not root successfully in open. There is limited success under constant growth regulators. Air layering with use of growth regulators is also reported to be of some success but did not reach commercial exploitation. Shield and patch methods of budding are practiced in Far East countries. It is not in vogue in our country. The method most commonly used in this country is therefore, inarching because of its high percentage of success. Inarching has some

Drawbacks:

- The rootstocks have to take to the scion tree for grafting.
- Scaffolding has to be built to hold the rootstock pots at scion level.
- The rootstock plants have to be watered individually while they are perched at various levels and at odd places hidden in the scion tree and all these increases the cost of production of grafts making it very cumbersome.

Several devices have been suggested to get over these difficulties:

- Grafting in rainy season, so that, the rains wet the stocks.
- Avoiding the pot of the rootstock by substituting it with grass wrapping.
- Use of grafting pot stands
- Reducing pre grafting and post grafting irrigation costs on the, by keeping the pots in a trench and letting in water to irrigate them.
- Training the scion trees to low heights.
- Wrapping roots in moss or soil and covering with polythene sheet so that no watering will be needed till union takes place

Amongst the grafting methods, the detached scion method is gaining popularity (veneer grafting).This has the striking advantage in using the scion sticks for propagation and the twigs from the variety to be propagated can be cut from the mother plant and taken to the rootstock seedling growing in a nursery. The scion stick, which is procured, must be 3-4 months old. It is found that scions ranging from 2.5 to 10 cm in length are suitable for the purpose. But the growth of the scion is always more when bigger scion pieces are used. If necessary the scion sticks can be stored for 6 days during April to June by wrapping them in moist paper and enclosing in an alkaline bag without any deterioration in the percentage of success. This is a less expensive and easy method of propagation to inarching.

Raising of Rootstock

The seeds (stones) are extracted from full ripe fruits within a week. The viability of seeds varies from 90-100 days. The seeds are sown directly in poly bags or raised beds with a mixture of farm yard manure, red soil and sand. The seeds are sown flat and germination occurs within 10-15 days. One year old seedlings are used for grafting.

Scion Selection

- Should be of pencil thickness
- Selected from healthy terminal shoots of 3-4 months age
- The scion should be defoliated 8-10 days before grafting. This is called procuring.

Rootstocks

Mangifera like *M. foetida* and *M. odorata* have been found to show promise as rootstock. Kalapad variety is said to be dwarfing stock and a wild mango variety Pulima (Ceylon) is considered as prolific rootstock. Polyembryonic seedlings which are uniform amongst themselves are recommended as rootstocks as they reduce variability due to the rootstock.

Sl. No.	Variety	Special characters
1.	Kurukkan	Salt resistant and Polyembryonic.
2.	Olour	Vigorous rootstock, Polyembryonic and dwarfing.
3.	Rumani, Kalepad and Vellaikolamban	Dwarfing.
4.	Moovandan and Nekkare	Salt tolerant.
5.	Pulima	Prolific rootstock

Inarching / Approach grafting

This method is extensively practiced in Tamil Nadu. Nine months to one year old rootstock is grafted with the scion which remains attached to the mother plant. In the rootstock, 20 cm above the ground level, 5 cm long slice of bark with wood is removed and corresponding cut is made in the scion so as to attach the cut surfaces firmly with jute thread. The tied portion is covered with cow dung mud paste. The graft is separated from the mother plant after 70-80 days of grafting.

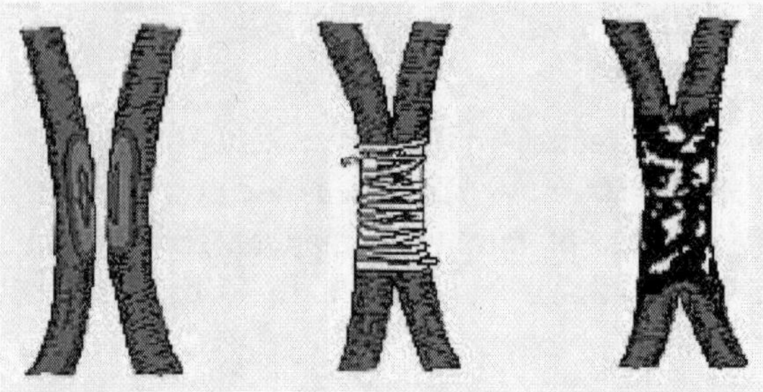

Fig. 2. Approach / inarch grafting

Epicotyl Grafting

In case of epicotyl grafting, tender seedlings with height of 15 cm are selected as rootstocks and a 'V' shaped cut is made after beheading it as a height of 4 – 6 cm from the cotyledons connectivity. The procured scion is collected and a wedge is made at the base of it, so as to exactly fit it in the cut made in the stock. The scion is exactly fitted in the stock and tied with polythene strips. The success of epicotyl grafting varies from 50 – 60 percent and depends upon high humidity, temperature, freedom from fungal disease, number of rainy days and rate of cambial growth.

Softwood Grafting

When the above method is adopted in 30 – 40 days old seedlings, it is known as softwood grafting. The success varies from 40 – 50 percent.

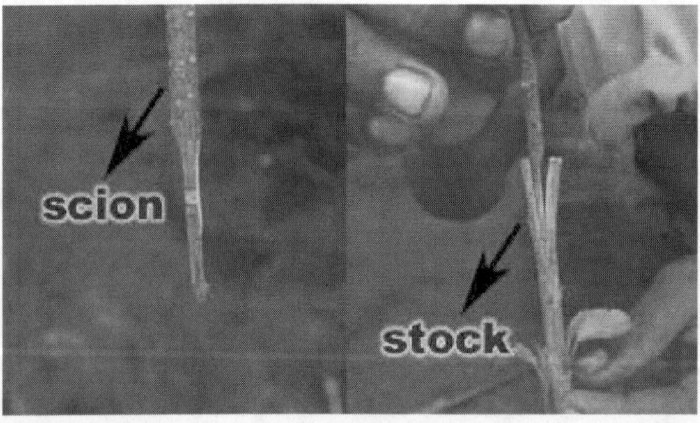

Fig. 3. Softwood grafting

Planting

Planting should be done during rainy season in pits of size 1x1x1 m^3. Different systems of planting like square, rectangular and hexagonal are followed but square system is most popular. In in-situ grafting, rootstocks are planted in the field and soft wood grafting is done in the following year during August – September. Normally spacing adopted for grafts is 10 X 10 m and for seedlings and in very fertile soils it is 12 X 12m. For dwarf varieties high density planting is considered. The spacing adopted under high density planting is 5x5 / 5X3 / 5X2.5 / 4x4 / 3x2.5 2.5x2.5m. While planting grafts, the graft joint or the union should be 20 cm above the soil surface to prevent entry of disease carrying organisms in to the graft joint. Immediately after planting plants are watered and staked. Under Tamil Nadu conditions, planting season varies from June to September. In north India, planting can be done in February to March if irrigation is available.

Planting Systems

Different systems of planting like square, rectangular and hexagonal are followed at different places. However, square and rectangular systems are also popular.

High Density Planting (HDP)

High density planting helps increase the yield/ unit area. In north India, mango Amrapali is found amenable for high density planting with a spacing of 2.5 m x 2.5. If coupled with pruning, it helps increase production/unit area in Dashehari. The polyembryonic mango Vellaikolumban when used as rootstock imparts dwarfing in Alphonso.

Advantages of HDP

- Higher yield and productivity per unit area.
- Effective utilization of land.
- Increase the efficiency of utilization of inputs such as fertilizers and water.
- Reduced labour cost.

Mango HDP

In North India, mango Amrapali is found amenable for high-density planting with a spacing of 2.5m x 2.5m.Normally for high density planting (HDP), the distance is 5 x 3 (or) 5 x 2.5 or 4 x 4 3 x 2.5 or 2.5 x 2.5m.Recently at the advent of many hybrid cultivars, a closer spacing of 2.5 x 2.5 to 5 x 5 m is recommended. In high density planting for Manjeera and Mahamooda the spacing adopted is 4.5 X 4.5 m and for Amrapali it is 2.5 x 2.5 m.

Variety	Spacing	No. of plants/ha.
Arka Aruna and Sindhu	5 m x 5 m	400
Amrapali	2.5 m x 2.5 m	1600

Ultra High Density Mango Plantation (UHDP)

This technology was developed by Jain Irrigation Systems Ltd., (JISL), Jalgaon, Maharashtra. Spacing is 3 m x 2 m and accommodates 674 Plants/Acre. Jain irrigation started large scale mango plantations during 1996 and adopted 4.5m x 4.5m, which is considered as medium density plantation (MDP). Results of MDP were remarkable and the orchard could start commercial production in the fourth year itself and later on yields were better than those from the conventional besides several other advantages.

Benefit to farmers

- Very low gestation period allows farmer to reduce cost burden and early income generation.
- Suitable for small, medium and semi-large farmers as it is management intensive and yields good quality fruits which can get better market price.
- It makes sensible bankable project to offer financial support due to assured high early returns.

Nutrient Management

- It is depends upon the variety, age, soil and irrigation.
- Follow INM (Integrated Nutrient Management).
- No inorganic manures in first year.
- At planting, apply 10 kg FYM/pit.
- Apply in September - October.
- 45-90 cm away from trunk up to peripheral leaf drip.
- All the chemical fertilizers should mix with FYM and apply.

Dose manures and fertilizers (Kg per tree)

Manures and Fertilizers	1 Year old	Annual increase	6th year onwards (Kg per tree)
FYM	10.00	10.00	50
N	0.20	0.20	1.0
P	0.20	0.20	1.0
k	0.30	0.30	1.5

Manures and fertilizers may be applied during September - October, 45 - 90 cm away from the trunk up to the peripheral leaf drip and incorporated. In Tamil Nadu the manures and fertilizers are applied in two split doses. First dose is given during June - July, consisting of complete Nitrogen with half potash at the onset of monsoon. The second dose is applied during September-October with FYM, super phosphate and remaining half of potash.

Foliar Application

Promotes growth, improves fruit set, yield and quality. Foliar spray is given immediately after the harvest of the crop. Spray urea @ 1-2% and another after flowering and at marble stage to improve fruit set and to reduce fruit drop.

Irrigation

The young plants upto 2-year-old should be watered regularly. The newly-planted grafts need about 30 litres of water every week. Irrigation during preflowering phase increases flowering. Irrigating grown-up trees after fruit set at 10-day interval increases the yield. However, for obtaining god flowering, the irrigation must be stopped atleast two to three months before the flowering period

Intercropping

In mango, intercropping helps check weed growth and reduces nutrient losses. Intercropping blackgram-wheat-mango and brinjal-onion-mango gives better monetary benefits. Besides, taking up cover crops like sunhemp, cowpea, pea help to prevent soil erosion.

Weed Management

- Root zone area of the trees must be kept weed free all the time.
- Manual cultivation is not recommended as it will disturb active roots.
- Weed killers with heavy mulch may be quite effective in controlling weeds.
- Weed control practices do not differ from what was explained for non bearing trees.
- Weeds can be controlled by the application of 4 kg/ha Atrazine/oxyflurofen (Goal) @ 800ml/ha as pre-emergence and application of 2 liters/ha Gramaxone (Paraquat)/as post emergence.

TRAINING AND PRUNING

Bearing Habit

It refers to the position of the fruit bud with respect to the vegetative growth. Mango is a terminal bearer in the past season's growth. The inflorescence is a panicle from terminal shoot and rarely from the auxiliary buds. Flowers are small, andromonoecious in nature with male and perfect flowers in the inflorescences. It is highly cross pollinated crop. The number of flowers varies from 1000-6000 per panicle. The flowering period is for 2-3 weeks and extends from November to February.

Training

Rootstock sprouts and low-lying branches have to be removed in the initial years of establishment. Besides, training is done to provide a good frame work for the future so that branches are spaced properly and these do not break with the crop load at the bearing stage. Thereafter, overlapping, intercrossed, diseased, dried and weak branches are to be removed after their harvest season is over.

Pruning

Mango is a terminal bearer in the past season's growth and normally does not require any pruning. But when the trees become very old, the bearing capacity decreases. This is due to imbalance in hormonal status, which might be due to excess vegetative growth and self-shading of most of lower branches. The productivity of old trees can be revived by judicial pruning practices. It involves heading back and thinning out.

Heading Back

Pruning is done during the month of August – September. In the heading back process, the hefty criss-cross branches are beheaded so that the centre is opened up. All the dead and dried branches should also be completely removed. By this, there will be abundant penetration of sunlight as well as aeration. The dormant buds in the old woods get activated. This should be practiced normally once in five years.

Thinning Out

In the thinning out process, the excessive terminal shoots all over the surfaces of the canopy are thinned out by retaining only one or two shoots for each previous season growth. This directs the nutrient and hormonal flow to restricted number of shoots, so that the terminal buds is converted into a reproductive bud. Thus the old unproductive trees can be made to bear a good crop. This has resulted in greater yield than in previous years in cultivar Mulgoa. Recently another method of pruning is recommend. It involves heading back of flowering shoots after harvest below a node. This helps to produce 3 to 5 new shoots below the pruned nodes which will bear the fruit buds during the nest season by November – December. This pruning should be followed by immediate irrigation and thereafter also at regular intervals to get desired effect.

Before thinning

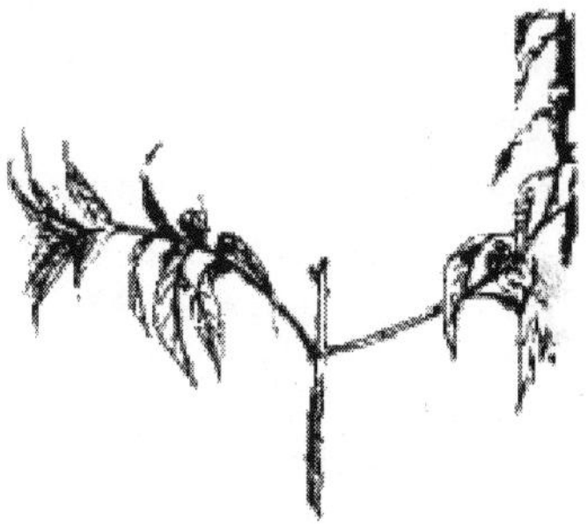

After thinning

Fig. 4. Thinning in mango

CLUSTER THINNING OF SHOOTS IN MANGO

Flowering and Fruiting

Mango is a terminal bearer i.e. fruits are borne on terminal shoots and almost all the varieties show alternate or irregular bearing habit which requires certain amount of shoot maturity before flowering. The vegetative growth is produced in two to five flushes-one during spring season, the next at the time of flowering land the subsequent ones depending upon favourable climatic conditions till the tree goes to dormant phase in autumn.

Though mango grafts start flowering in the second year, crop should be taken only after the 4th or 5th year. The grafted mangoes come into bearing in five years and commercial production can be obtained in about seven years. Seedlings take about 10 years. The precocious varieties like dashehari start bearing even from the third or fourth year onwards. A grafted mango tree usually continue to bear till up to 50 years, but its most prolific years of bearing are generally from the 10 to 14 years. Seedling trees usually take about 8 years to come to bearing but their productive life is much longer.

Flowering time is closely linked with the fruit-bud differentiation which varies with the variety and the area where it is grown. In most cases, fruit bud differentiation occurs between October and December. The flowering in mango takes place as early as November – December in Tamil Nadu, February to March in north India and January - February in eastern parts of the country. Off-season flowering in Tamil Nadu is during August - September.

Regular cropping can be obtained by planting regular bearing varieties such as Totapuri, Neelum etc. (or) Five sprayings of Ethrel at a concentration of 200 ppm coupled with 0.1 % urea are given, beginning from middle of September at monthly intervals and deblossoming the trees in 'on' years.

Mango inflorescence consists of two types of flowers, viz. male and perfect flowers. The number of flowers per panicle varies between 1000 - 6000, depending upon the variety. The proportion of perfect flowers determines the initial fruit set in mango, but final retention depends on the climatic conditions, Varietal characteristic, etc. Mango is a highly crosspollinated crop and pollination is through insects (Main pollinator is house fly [*Musca domestica*]) and flies.

Successful pollination requires a dry and frost free period. During rains and dew, insect activity is reduced, the pollen becomes sticky, and increasing the incidence of pests and diseases and the resultant effect impedes pollination and fertilization. Commercial mango varieties such as Dashehari, Chausa, Langra and Bombay Green are self-incompatible and hence Bombay Green for Dashehari and Dashehari for Chausa should be used as pollinizers.

Bearing Habit

- It refers to the position of the fruit bud with respect to the vegetative growth.
- Mango is a terminal bearer in the past season's growth.

Flowering Trait

- The inflorescence is a panicle from terminal shoot and rarely from the axillary buds.
- Flowers are small, polygamous, monoecious or hermaphrodite, protogynous in nature.
- Only 0.1 % flowers (perfect) develops fruits to maturity.
- It is highly cross pollinated crop. The numbers of flowers varies from 1000 - 6000 per panicle.
- The flowering period is for 2-3 weeks and extends from November to February.
- The duration of flowering is very short, usually of 2 to 3 weeks.

Use of Growth Regulators and Chemicals

To Induce Flowering

Soil application of Paclobutrazol (Cultar) @ 5 ml/10 litre of water/tree for full bearing tree during August - September. Salicylic acid 2000 ppm - successful in Amrapali and Dashehari. Malformation was minimum. Spray urea 0.5% (5 g/lit.) (or) 1% Potassium Nitrate (KNO_3) (10 g/lit) during February. It may spray to induce flowering, if trees do not flower by that time.

To increase the fruit retention and to control fruit drop: Two sprays of NAA 20 ppm (Planofix 1 ml/45 lit). (One at flowering and the other at marble stage). Spray 2% Potassium Nitrate (KNO_3) at mustard size to increase fruit set and retention of fruits.

USE OF PACLOBUTRAZOL (CULTAR) ON MANGO FLOWERING

A. To induce Regular Flowering and to Overcome Alternate Bearing

Soil application of Paclobutrazol (Cultar) @ 5 ml/10 litre of water/tree for full bearing tree during August - September. Application of Paclobutrazol @ 3ml/m canopy diameter at 90 120 days before bud brust is recommended for increasing the yield and to control alternate bearing in mango cultivars. Apply Paclobutrazol @ 10 g a.i. / full bearing tree during first fortnight of September to get maximum number of fruits and yield during off years.

To induce off-season flowering (August - September):

- Soil application of Paclobutrazol (Cultar) @ 5 ml/10 litre of water/tree during April - May.
- The flowering will occur within 90 to 120 days.
- Heading back of crowded branches and centre opening with use of Paclobutrazol is overcrowded mango orchard is recommended for commercial adoption. Retardant i.e. Paclobutrazol (pp333) is recommended to convert "off year" into "on year" Application of pp333 is done during 15th July to 31st August.
- Soil drenching with paclobutrazol (2ml/ tree) induces flowering during off year. It has become a commercial practice in Konkan region of Maharashtra.

Harvesting Indices or Maturity index

It requires considerable experience to judge the stage of maturity at which the fruits are to be picked. If the fruits are harvested at an immature stage, they develop white patches and the normal TSS / Acid ratio is also not obtained. On the other hand, if the fruits are harvested when over mature or fully ripe, more microbial and physiological spoilage occurs due to increased susceptibility of fruits, thus reduces the shelf life of fruits. The correct stage of harvesting of fruits can be judged in several ways.

- Attainment of full size of a variety is indicated by stoppage of increase in size and flesh becomes soft.
- At maturity, the colour changes from green to yellow and the development of "shoulders" on the stem end of the fruit are the best indicators of maturity.
- Flowering to maturity - 110 - 130 days.
- Development of waxy coating (Eg. Neelum) and red blush (Senthura and bangalora) on matured fruits.

Harvest

The fruit injured or even slightly bruised during the picking operation losses its keeping quality and becomes unfit for dispatch to distant markets. The usual practice of harvesting the fruits is knocking down the fruits or shaking the trees violently to get the fruit down need to be discouraged. If the fruit is hanging high, it should be picked up with the help of a step ladder. In no case, the picker should climb up the tree as this practice is harmful both for the tree and the picker. The fruits should always be harvested and the harvested fruits are placed in canvass bags or padded baskets and carried to the packing house.

Yield

The yield of mango depends on no. of factors like age of the plant, soil fertility, climatic conditions, variety, type of plant material like graft or seedling, management of the orchard etc. A mango tree having an age of 5 years bears about 200 fruits and 10-20 years bears a crop of 500-1000 fruits during an year. During 20-40 years its yield may range between 1000-3000 fruits.

Yield Varies with Varieties and Spacing Adopted

- The average yield in our country is about 7.5 to 10 tonnes per hectare.
- 8-10 t/ha - Regular varieties.
- 15-25 t/ha - Hybrids.
- 400-600 fruits/year/tree (10 year old tree).
- The number of fruits per tree during its steady bearing age (i.e., after 20 years) generally varies from 1000 to 2000 fruits.

POST-HARVEST TECHNOLOGY (PHT)

Grading

After harvesting, mangoes are graded according to their size.

According to weight

- Grade 1 - 200 g and above.
- Grade 2 - 170 to 200 g.
- Grade 3 - Less than 170 g.

Packing

Fruits can be washed prior to packing to remove surface resins and then dipped in hot water (55°C). To maintain the quality, proper packaging is a must. In western region, bamboo baskets are used for packing. A basket contains 50-100 fruits. Straw is used for packing. Wooden boxes are also used in some place. However, now perforated cardboard are generally used. In these boxes either fruits are individually wrapped with tissue paper before packing or paper shavings are used for cushioning.

Storage

Fruits are stored for 15 days at 15°C and RH (Relative Humidity) of 85-90%. Ethylene is used to accelerate color development (by 3-8 days) and allow more uniform ripening. Dip the fruits in 52.1°C hot water immediately after harvest for 5 minutes followed by 8% plant wax (Fruitox or Waxol) to reduce anthracnose disease in mango during storage.

Two pre-harvest sprays of 0.20% Mancozeb (2.0 g/lit) will also reduce the incidence. Mangoes are susceptible to chilling injury and must not be stored at below 10 - 12°C. Storage life is only 2 - 3 weeks under optimal conditions. Hydro-cooling at 12 - 15°C and holding for 2 weeks at 15°C followed by storage for 1 week at ambient temperature gives good storage life to fruits.

PHYSIOLOGICAL DISORDERS - PROBLEMS IN MANGO CULTIVATION

1. Alternate (Biennial) and Irregular Bearing in Mango

It is a burning problem in mango industry since; it renders cultivation of mango less remunerative to the growers. When the mango tree bears heavy crop, the bearing season is called 'on' year. When it bears a poor or lean or no crop, the season is called 'off' year. In both the years the mango grower is at a loss. In the 'on year' there is abundant fruiting and the market is flooded with fruits and the prices fall down. In the 'off year', there is scarcity of fruits for supply to market, prices are highly remunerative but the grower has a very little or no fruits to supply to the market. This phenomenon is called 'irregular bearing' or 'biennial bearing'. Most of the promising varieties like Alpphonso, Dashehari, Langda, Mulgoa, Kesar etc. Various factors are reported to be responsible for this phenomenon. These phenomena are mainly due to specificity of varieties, climatic changes in a locality, imbalance in soil nutrition status and consequent changes in balance of these nutrients in the shoots of the tree mainly the carbohydrate / nitrogen ratio and hormonal substances. No single reason could be pin pointed to this strange phenomenon.

Causes

Internal factors: Physiological, genetical, nutritional, sex ratio and hormonal imbalance.

External factors: Rain, wind, low temperatures, cloudy weather, light, pests and diseases.

Varietal Factors

The habit of bearing heavy and poor or no crop in alternate years is believed to be an inherent character because some varieties, viz., Fazil, Neelum, and Bangalora are regular bearers.

Maturity and Age of Trees and Seasonal Vegetative Growth

In mango the maturity, age of trees and seasonal vegetative growth influence and control the flowering. In mango a definite relationship appears to exist between growth of vegetative flushes and fruit bud formation. An early production of shoots during the first flush (February-April), and an early cessation of the growth in the season seem to be conducive to regular bearing in mango. But in several old and neglected orchards the tree bearing heavily in a year put forth little or no vegetative growth during flowering to fruit maturity, and as such no fruiting wood is available for the following year, which consequently, turns out to be an off year for cropping.

C: N Ratio and Nutritional deficiency

It is also contended that irregular bearing in mango was caused by nutritional deficiency especially Nitrogen. A proportionate increase in nitrogen leads to vegetative growth, whereas its proportionate decrease induces flowering. Higher starch reserves, total carbohydrates and C: N ratio favour flower bud formation but not in Baramasi and regular bearing varieties. Studies on nitrogen content in stems and leaves of different cultivars indicated possibility of high no. of trees to flower, but no correlation between flower bud initiation and total nitrogen could be derived. Thus C: N ratio reserve, though playing an important role in flowering, is not the primary cause of biennial bearing.

Hormonal Imbalance

Higher level of auxin like substances and an inhibitor (similar to ABA) and lower levels of gibberellins (GA_3) like substances are vital for floriferous shoot in mango.

Inflorescence Type

The varieties that produce mostly terminal inflorescence and only few axillary types of flower clusters are more markedly biennial bearers; while those varieties which produce a greater percentage of axillary inflorescence are moderately regular bearing. Unfortunately, most of our good dessert varieties are biennial bearers.

Climatic Factors

Adverse climatic factors such as cloudy weather, rains, high humidity, convert an 'on year' in to an 'off year' by promoting incidence mango hoppers and diseases like powdery mildew and anthracnose etc., especially during flowering, preventing pollination and damaging the floral parts. Frost and low temperatures during flowering adversely affect fruit set there by turning an 'on year' in to 'off year'. However, these don't form basic causes of biennial bearing. In view of the above various possible factors associated with this problem can be grouped as internal factors and external factors.

Suggested Measures to Overcome Alternate Bearing (or) Control of Alternate Bearing

- Selection of regular bearing varieties specific to a locality. Neelum and Bangalore are regular bearers under South Indian conditions which are not much affected by changes in weather conditions in a particular year.

- Planting of varieties which are regular bearers Eg. Bangalora, Rumani, Banglora, Neelum, Mallika, Amrapali and Ratna.
- Plough and harrow the orchard twice in a year, in the beginning of the monsoon after
- harvest and in October.
- Apply the fertilizers as per recommended schedule.
- After the harvest of the crop, dead and diseased branches and the Loranthus parasite should be removed.
- Providing orchards with wind breaks, regular ploughing, and liberal manuring at appropriate intervals and adequate irrigation and other cultural practices are also recommended for regular bearing.

Deblossoming

Half the no. of flower clusters are removed from the tree in the on year as soon as they emerge. The food reserves of these deblossomed shoots would be utilized by the tree in producing vegetative growths in the spring and summer and mature to produce.

Over vigorous trees may be subjected to induce rest by withholding irrigation with good exposure to sunlight. Application of common salt @ 2.5 kg. Per tree in September checks vegetative growth causing accumulation of carbohydrates for induction of flowering. This is practiced in over-vigorous trees and not on weak trees.

Pruning

Proper pruning of mango trees after the harvest gives the best results in terms of disease and pest management, diversion of food materials to the productive shoots, increased photosynthetic activity and increased carbohydrate and starch content, early production of new flush that bear the crop in the next season. Pruning also helps in increased cytokinin content, Ascorbic acid and auxin, which was beneficial for flowering. Abscissic acid level will also increase which inhibits the vegetative growth and promoted flowering. Thus pruning helps in obtaining regular fruiting and production of quality fruits. Smudging during October-December is reported to be useful in forcing Mango to flower with limited practical utility.

2. MANGO MALFORMATION

Most of the flowers lack essential organs and don't set fruit and its inflorescence continuously hang on the tree for months, being more green and sturdy. It is a complex and serious malady in North India and not so common in South India. Some varieties like Chausa, and Bombay green are more susceptible than others in N.India. It is one of the most important disorders, causing huge losses. It is a major problem in Punjab, Delhi and Uttar Pradesh. However, it has also been noticed in Gujarat, Maharashtra, Bihar, West Bengal and Orissa.

There are the 2 types of mango malformation, they are, vegetative malformation and floral malformation.

Vegetative Malformation

It is more common in nursery seedlings and young plants. Affected mango seedlings or young plants develop excessive vegetative branches which have limited growth, swollen and with very short internodes and develop abnormally compact rosette like shoots presenting a bunchy top appearance.

Floral Malformation

Floral malformation affects trees at the bearing stage. Directly affects the productivity. The affected inflorescence becomes clustered. The incidence of disorder varies from variety to variety. Various causes like nutritional disorders, physiological, viral, fungal and acarological are reported. The available literature doesn't appear to support the above causes except the fungal; the only positive evidence for the cause of this is fungal origin *i.e., Fusarium moniliforis.*

Control Measures

- Destruct the infected small plant or plant parts Spraying a fungicide like Captan @ 3grams or Bavistin @ 1gram / litre of water.
- Remove the affected parts by pruning 30cms below and paste with Bordeaux paste.
- Early deblossoming combined with NAA 200ppm spray during October also reduces this considerably.
- Bhardhan variety of U.P. is resistant and free from this infestation.

3. SPONGY TISSUE

Fruits from outside look normal but inside a patch of flesh become spongy, yellowish and sour. A non-edible patch of flesh develops in the mesocarp of the fruit and becomes spongy, sour and yellowish is termed as spongy tissue. This can be detected only after cutting the ripe fruit. It is a physiological disorder in which the fruit pulp remains unripe because of unhydrolised starch due to inactivation of ripening enzyme because of high temperature, convective heat, and post-harvest exposure to sunlight are the causes. Alphonso variety is very susceptible to this spongy tissue.

Control

Sod culture (growing grasses in orchard) and mulching are useful in reducing spongy tissue. Growing mango hybrids Ratna and Arka Puneet, which are free from this problem, have Alphonso like characters and do not suffer from this malady. Harvesting fruits when they are three fourths matured rather than fully matured ones also reduces this malady.

4. FRUIT DROP

Fruit drop is serious problem in mango and cause great loss to the growers. A tree producing several thousand panicles yields only a few hundred fruits. Most of the flowers falling down after full bloom or at later stage of development. Only 0.1 to 0.25% perfect flowers or even less develop in to mature fruit. Maximum fruit drop takes place in last week of April or first week of May depends upon favourable condition. The fruit drop can be divided in to three distinct phases Eg. Pin head drop, post setting drop and May drop. The flower drop as well as the fruit drop is primarily due to the formation of an abscission layer at the point of attachment of the fruit with the twig. Several factors have been considered responsible for the formation of abscission layer. The causes can be divided in to two;

External Causes

- Unfavourable climatic conditions.
- High incidence of serious diseases like powery mildew and anthracnose and pests like hoppers and mealy bugs.

Internal Causes

- Poor soil.
- Lack of pollination.
- Low stigmatic receptivity.
- Defective perfect flowers.
- Poor pollen transference.

- Occurrence and extent of self incompatibility.
- Abortion of embryo.
- Degeneration of ovules.
- Competition between developing fruit lets.
- Drought / lack of irrigation.

Measures to Prevent Fruit Drop

The extent of fruits drop can be reduced significantly by regular irrigation during the fruit development period, maintaining sufficient soil moisture also prevents fruit drop and helps in increasing the size of the fruit and timely and effective control measures against major pests and diseases.

Spraying of 2, 4-D @ 10 ppm or NAA @ 50 ppm at pea stage and at marble stage helps in preventing fruit drop. Through the application of growth regulators like NAA (50 ppm) and 2, 4-D (20 ppm) during off years about six weeks after fruit set. Providing pollenisers for self incompatible types. Provision of wind beaks all around the orchards, which prevents drop due to high velocity winds at the time of fruit development.

5. BLACK TIP

This disorder is mainly noticed in Punjab, Uttar Pradesh, Bihar and West Bengal. The distal-end of the affected fruits turns black and becomes hard. These fruits ripen prematurely and become unmarketable. This disorder is caused by the smoke of brick-kilns located within a distance of 1.5 to 2 km. Gases like carbon monoxide and carbondioxide, sulphur dioxide and acetylene cause these symptoms. It can be controlled by raising the height of the chimney of the brick-kilns. Spraying borax (0.6%) and caustic soda (0.8%) at 10-14 days intervals starting from before flowering, flowering and fruit set satge can also controls it.

6. CLUSTERING DISORDER IN MANGO (JHUMKA)

A fruiting disorder, locallly known as 'Jhumka', is characterised by the development of fruitlets in clusters at the tip of the panicles. Such fruits do not grow beyond pea or marble stage and drop down after a month or so of fruit set. These fruits do not contain seeds when they are cut open. The disorder seems to be due to lack of pollination / fertilization which may be attributed to many reasons. Among them, absence of sufficient population of pollinators in the orchards is the major reason.

Some of the remedial measures are suggested below:

- Insecticides should not be sprayed at full bloom to avoid killing of pollinators.
- Pests and diseases should be controlled in time by spraying the recommended
- pesticides only.
- Introduction of beehives in the orchards during flowering season for increasing the number of pollinators.
- The practice of monoculture of a particular variety may be avoided. In case of Dashehari, 5-6 per cent of other varieties should be planted in new plantations. In old orchards, where monoculture of a particular variety like Dashehari is followed, a few branches may be top worked with pollinizing varieties.
- Pruning of old trees may be done to open the canopy.
- Spraying of 300 ppm NAA may be done during October-November

❑❑❑

CHAPTER

8

BANANA

Botanical name: *Musa sp.* L.

Family: Musaceae

Chromosome no.: 2n = 22, 33 & 44

Origin: South East Asia

The banana is considered to be one of the most primitive, cultivated tropical fruits in India. It is the second most important fruit crop of India. Banana is one of the oldest fruit known to mankind and also important food for man. It is one of the oldest and commonest of the Indian fruit that has been cultivated since ancient times. Its antiquity can be traced back to Garden of paradise where eve was said to have used its leaves to cover her modesty in the garden of paradise. It may be one of the reasons why banana is called "Apple of Paradise" and botanically named *Musa paradisiaca.* Banana plants refer to Biblical legend as "Tree of wisdom" for good and evil in the Garden of Eden. Banana by virtue of its, multiple uses is popularly known as "Kalpataru" (a plant with virtue). It is used as staple fruit in most of the African countries and is used as ripe (table) or raw fruit (cooking). Edible bananas are mostly hybrids of the two species from *M. acuminata* and *M. balbisiana.* They set fruits by parthenocarpy. It is called as 'Apple of paradise' and 'Adam's fig'. This is the fruit for all ages and it is a good laxative. It is one of the fruit among 'Mukkani' or Triplet fruits, comes last (Ma-pala-vazhai). It is equally suitable both for large scale cultivation as well as for home scale cultivation.

Nutritive Value

Banana is a rich source of energy in the form of sugars and starch. Fruit for all ages. Banana is eaten when ripe as dessert / table fruit. It is good laxative. Rich source of carbohydrates, Vitamins, riboflavin, niacin, Ascorbic acid and Minerals -Ca, Mg, K & P. Rich source of energy (137 K. Calories/100g) or (350 to 550 kilo joules/100g).

Climate

The banana is strictly a tropical crop. It grows luxuriantly in the warm, humid and rainy climate of tropical regions of the equator. It grows well in regions with a temperature range of 10-40°C and an average of 23°C. In cooler climate the duration is extended, sucker production is affected and bunches are small. Low temperature i.e. less than 10°C is unsuitable since, they lead to a condition called "choke throat" or impeded inflorescence and bunch development. Banana grows well under high rainfall areas .On an average 100 mm rainfall per month appears to be satisfactory for growth of banana. Hot winds blowing in high speed during the summer months shred and desiccate the leaves. Stagnation of water is injurious and may cause diseases like panama wilt. Banana comes up well up to an altitude of 1500m above sea level in the tropics. The hill bananas in Tamilnadu are raised between elevations of 500-1500m mostly under rain fed conditions. Absence of strong winds and cyclones are important.

Soils

Banana is voracious feeder and requires a well-drained soil with plentiful organic matter. Even though banana requires heavy irrigation, it cannot with stand water stagnation. Therefore, the soil should be well drained and deep (At least 1m depth).It grows successfully in loamy soils, well drained clay soils of delta areas, irrigated medium soils. The production in lighter soils is good. Saline soils with salinity percentage exceeding 0.05 are unsuitable. Banana can grow well even under slightly alkaline soils. Such soils are found preferable for avoiding the wilt disease, which is known to be severe in acid soils.

Varieties

Most of the present day edible banana varieties have originated from the two species *viz., Musa acuminate* and *Musa balbisiana.*

The notation of 'genome' was 'A' for acuminata and 'B' for balbisiana as suggested by Simmonds and Shepherd (1955) as a key to the classification. 'A' represent, acuminata genome with 11 chromosomes from *M.acuminata*. While 'B' represents genome with 11 chromosome from *M. balbisiana.*

Names of banana varieties are often confused by local synonyms.

- AA - Matti, Sana Chenkadali, Namarai (Diploid acuminata types)
- AAA - Gross Michel, Grand Naine, Dwarf Cavendish, Robusta & Red banana (Triploid acuminata types)
- AAAA - Bodles Alta fort (Tetraploid acuminata)
- AB - Ney poovan, Kunnan (Diploid mixed)
- AAB - Poovan, Rasthali, Hill Banana, Ladan, Nendran (Triploid mixed)
- ABB - Monthan, Karpura valli (Triploid mixed)
- ABBB - Klue taperod (Tetraploid mixed)

Matti (AA): It is commonly grown in Kanyakumari district. Fruits are small – taste is good. Good male parent.

SannachenKadali (AA): It resembles red banana. It performs well even under full shade. Hence, it can be grown in coconut gardens.

Namarai (AA): Grown in Kanyakumari district. Good male parent – produce fertile pollen grains.

Gross Michel (AAA): Has desirable fruit characters. It is susceptible to wilt. Now not grown commercially.

Dwarf Cavendish (AAA) suitable for wind prone areas.

Synonyms: Basrai Dwarf, Pachavazhai, Mauritius, KullaVazhai, Kuzhi Vazhai, Kuttavazhai.

a. Plant stature is dwarf.
b. Crop duration is 10-12 months
c. Each bunch weight 18-20 kg.
d. Fully ripped fruits will be greenish yellow in colour.

Robusta (AAA): Variety with big fruits and heavy bunches.

Synonyms: Bombay Green, Harichal, Pedda Pacha Arati.

a. Semi tall sport of Dwarf Cavendish.
b. Average bunch weight is 40 kg.

Red Banana (AAA): Medi banana with medicinal properties.

Synonyms: Lalkela, Chenkadali, Sevvazhai Rathambala.

a. The colour of the pseudostem, petiole midrib and fruit is purplish red.
b. Bunch weight 20-25 kg with 100 fruits.
c. Crop duration from planting to harvest: 18 months.

Ney Poovan (AB): Small fruited.

Synonyms: Rasakadali, Elakki, Senoreeta, Nallippovan.

a. Rind of the fruit is very thin and papery with yellow colour.
b. Bunch weight – 12kg with 150 fruits/bunch
c. Duration : 13 months.

Kunnan (AB): Small fruited

a. Quality variety of Kerala
b. Fruits with firm pulp.

Poovan (AAB):

Synonym: Champa, Lalvelchi, Mysore, Palayam kodan, Fill basket.

a. Individual fruit has prominent nipple.
b. Average bunch wt. – 15 kg having 8-12 hands each hand with 11-18 fingers.
c. Crop duration - from 11 to 14 months.

Rasthali (AAB):

Synonym: Silk fig, Martaman, Malbhog, Amruthapani, Rasabale, Vellai thuzhuvan.

a. Plant is tall and can be identified by the yellowish green pseudo stem with brownish blotches, reddish margins of the petiole and leaf sheath.
b. Bunch wt. – 12 kg. Fruit peel is thin with silk flesh.
c. Easy dropping of ripe fruits from the bunch (Poor pedicle attachment)
d. Formation of hard lump in the pulp.
e. Duration 14 months.

Ayirankai Rasthali (AAB):

Synonym: Ayirankaichi

It is a natural mutant of Rasthali. The fruits are longer at basal end and become smaller towards tip. No male phase and number of fruits ranged from 600-700 per bunch. Duration: 15 ½ months.

Hill Banana (AAB):

Synonym: Virupakshi, Sirumalai, Malavazhai, Vella Vazhai.

a. Perenial banana liked for its good flavour and keeping quantity.
b. Bunch weight 10-12 kg with 7 hands.
c. Duration 14 months.

d. Each fruit has 5 ridges/sides.

e. Peel is thick. It has 5 ridges/sides.

f. It peels off easily from the pulp.

Ladan (AAB):

Synonym: Nadan, Singan.

g. Variety with large leaves can be grown under shade.

h. Stem is greenish with purple tinge.

i. Duration – 12 ½ months.

Nendran (AAB)

Synonym: Ethankai, French Plantain.

a. Pseudostem has a distinct shade of pink colouration.

b. Fruit is longer, peel is thick.

c. Bunch wt. – 12-15 kg.

d. Fruit with 3 prominent ridges strongly held.

e. Duration : 11 to 12 months.

Monthan (ABB):

Synonym: Bontha, Kanch kela, Bluegoe.

a. Culinary banana

b. Duration 12-14 months.

c. Bunch wt. 20-25kg with 80-100 fruits per bunch.

Sambarani Monthan (ABB)

A mutant of monthan with fruits having an ashy coated rind.

Karpoora valli (ABB)

Synonym Raja vazhai, Karpura vazhai, Pisang Awak.

a. Pseudo stem is light green with purplish tinge. Huge and tall

b. Peel yellow with ashy coating.

c. Pulp is very sweet with pleasure flavour.

Naatu Vazhai (ABB)

Synonym: Vayal Vazhai.

d. Popular in Southern districts of TN – unripe fruits are used for culinary purpose.

e. Ripe fruits for dessert purpose. Good coolent.

CO-1 (Hybrid) from TNAU, Coimbatore (1983)

It is a multiple hybrid developed from

Ladan (AAB) x *Musa balbisiana* clone sawai (BB)

↓

F_1 (AB) x Kadali (AA)

↓

Co.1 Banana (AAB)

a. The fruits have flavour and taste similar to Hill banana, at the same time, the plants can be grown in plains.
b. Bunch weight 10.5kg having 7 hands with 80-85 fruits / bunch. Each fruit weight about 150g.
c. TSS 22.6° Brix. Crop duration is 14-15 months.
d. The fruits will be sweet only after ripening.

Hybrids

BRS – 1 (Agniswar x Pisang Lilin) : Early ratooning ability, 100 days earlier then Rasthali, resistance to sigatoka leasf spot, (KAU, Kerala)

BRS – 2 (Vannan x Pisang Lilin) : Dark green poovan like fruits, 14 kg of bunch weight, it is having 8 hands, number of fruit should be 118 and resistance to sigatoka leaf spot, rhizome weevil and nematodes, (KAU, Kerala).

Gold Finger (FHIA01)

It is a hybrid between Dwarf Prata x SH3142 developed at Honduran federation of Agricultural Research. It is adapted to different soils – fruits ripen slowly. It could replace Cavendish banana. Bunch wt. 20-25kg.

Propagation

Banana is propagated vegetaively through suckers and rhizomes.

Suckers

There are two types of suckers. 1) Water sucker 2) Sword sucker.

1. **Water sucker:** Have broad leaves and broad pseudostem and they don‘t produce a healthy banana clump and hence not suitable for planting.

2. **Sword sucker:** It has a strong large base, gradually tapering to a slender point with one or two narrow sword like leaves at the tip. The sword sucker is most vigorous, grows fast and comes in to bearing early. Hence, sword suckers are preferred for planting.

Rhizomes

After harvest, a number of its suckers are encouraged to grow up to 1-2 feet. They are then dug out and their pseudostems are completely cut above the solid rhizome and roots removed. The rhizomes weighing about 450-900 grams are stored for two months in a dry place under shade. During storing the bottom remains cut off leaving the heart bud prominent at the top of the rhizome. The conical rhizomes, which have a sound heart, will be selected for planting. Some times the rhizomes are cut in bits of 1 kilogram and are used for planting.

Selection of Sucker

Select only 3-4 months old suckers from healthy vigorously growing and good yielding plants. The sword suckers should have 1.5- 2 kg weight.

Preparation of the Sucker

The selected suckers should be separated from its mother plant along with a portion of a rhizome. Later, the stem of the suckers should be beheaded at 20-30 cm height in a slanting manner. This helps in producing new leaves quickly. The slanting cut also prevents the stagnation of water in the sucker. The old roots should be removed and then dipped in 0.1% cereson @ 1 g.per litre of water for 5 minutes before planting.

Selection of Planting Material

- Sword suckers should be vigorously growing and weighing 1.5 to 2.0 kg
- The suckers should be swollen at the base and tapering towards the top.
- Free from pests, diseases and nematodes.
- They are prepared by trimming off the roots and decayed portion of the corm and cutting the pseudostem leaving 20 cm from the corm.
- The suckers can be graded and grouped based on their size before planting.

Pre treatment of suckers (Paring and Pralinage)

- The rhizomes selected from mother plant should be free from pests, diseases and nematodes.The rhizomes are trimmed and then dipped for 5 minutes in 0.2% Carbendazim solution (2g in 1 lit of water). This treatment will help to control Panama wilt.

- Then the sucker is dipped in clay slurry solution of 4 parts clay plus 5 parts water to have a thick coating of slurry over which 40g of carbofuran 3G granules is sprinkled. This treatment will control nematodes.
- Alternatively, the corms can also be dipped in with 0.2% Monocrotophos (2ml Monocrotophos 36 WSC / 1 of water), shade dried for atleast 24 hours.The sucker is then planted at the centre of the pit and the soil is pressed around.

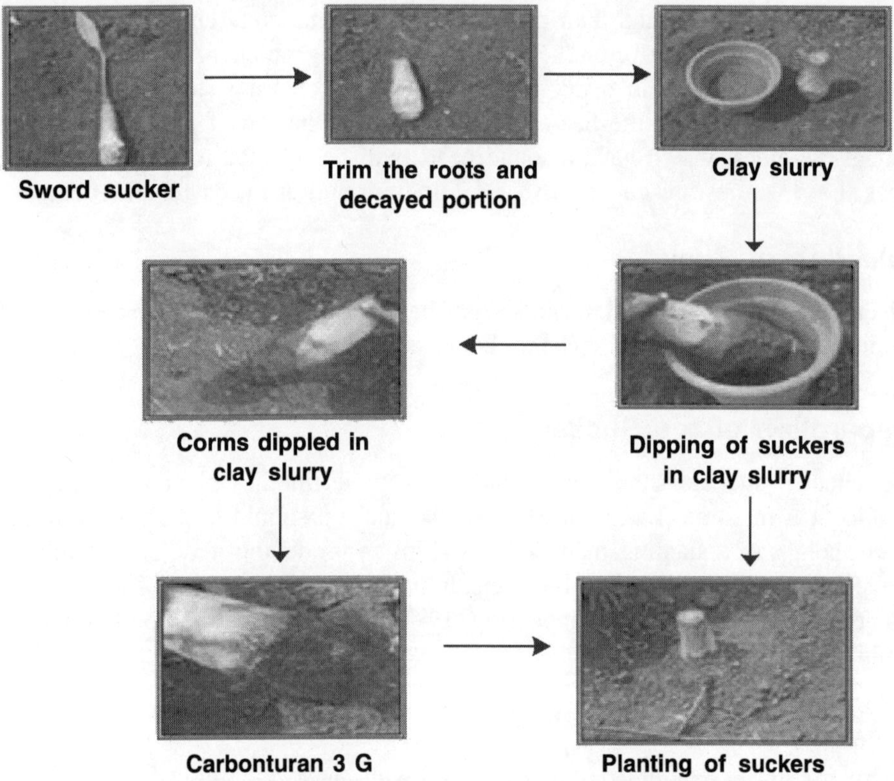

Fig. 5. Sucker treatment and planting

Land Preparation

The land should be deeply ploughed, harrowed and leveled and pits of 45cm x 45cm x 45cm should be dug at the required spacing. The pits should be exposed for weathering for about a week to control the presence of any grubs, ants, weevils etc. The soil from the pits should be mixed with the following thoroughly.

- 5-10 kg of FYM
- 0.5 kg of castor or neem cake

- 2 kg of wood ash or 50 grams of muriate of potash
- 200 grams of super phosphate.

The pits are then filled with the above-manured soil thoroughly. While filling the pits with the manured soil apply to the pit 50 grams of lindane dust to control weevil which affects the rhizome or sucker in the soil. The sword suckers are later planted straight in the pit along with a portion of rhizome at a depth of 10-15cm.

Planting

The period of planting should be such that the active growth phase of the plant may continue un-hampered during the flower bud initiation stage or stage at which embryonic bunch is formed inside the pseudostem. This generally occurs between 4 and 5 months after planting. This stage determines the no. of hands / fingers in future bunch after planting. At this stage there should be any extreme cold or hot weather or lack of soil moisture or lack of nutrients in the soil. June-July (On set of monsoon) is the planting season.

In general the beginning of monsoon i.e. June is the best time for planting banana in most parts, as the rapid growth during first 4 months of monsoon is particularly helpful. In the heavy rainfall tracts like Malabar planting is done after the cessation of monsoon from September to October. This also makes the plants quite small during the expected periods of high winds, storms and cyclones etc.

System of Cultivation in Banana

System of cultivation	Preparatory cultivation	Variety	Season of planting	Spacing
Garden land	2-3 ploughings, then taking pits and farming channels and basins	Robusta Nendran, Dwarf Gavendish, Grand Naine	Jan – Feb and Nov – Dec	1.8 x 1.8 m 1.5 x 1.5 m
Wet land	No elaborate preparation is required since banana follows paddy. Planting is done in small pits, sufficient to accommodate the corms	Poovan, Rasthali, Monthan, Neyvannan	Feb – April	2.1 x 2.1 m
Perennial (Padugai)	One deep mammutti digging, annual deepening of the water channels and spreading of the silt in the bunds.	Poovan, Monthan	Jan – Feb Aug – Sept	3.3 x 3.3 m
Hill banana	Scrub jungles are removed, contour walls are constructed.	Virupakshi, Sirumalai	Apr – May Jun – Aug	3.6 x 3.6 m

Application of fertilizers

Particulars	N	P	K
	(g / plant / year)		
Garden land			
Nendran	150	90	300
Other than Nendran	110	35	330
Wet land			
Nendran	210	35	450
Rasthali	210	50	390
Pooven and Robusta	160	50	390

Hill Banana

- 375 g of 40:30:40 NPK mixture and 130 g MOP/clump per application during October, January and April. *Azospirillum* and *Phosphobacteria* - 20 g each at planting and 5th month after planting preceding chemical fertilizer application.
- Apply N as neem coated urea. N & K in 3 splits 3rd, 5th and 7th month P at 3rd month of planting.
- For tissue culture banana 50% extra fertilizer at 2nd, 4th and 6th and 8th month after planting.

Irrigation

Irrigation should be given immediately after planting and life irrigation after 3-4 days; subsequently irrigations are to be given once in a week for irrigated plantations of garden lands and once in 10-15 days for wet lands. The field should be sufficiently irrigated after every manuring.

Drip irrigation can be advantageous to reduce water loss in conveyance and improving efficiency. Drip systems are especially useful in salt affected soil. Nearly 40 % savings in water is noticed when drip systems are used. 15 lit of water / plant / day from planting to 4th month, 20 lit / plant / day from 5th to shooting and 25 lit / plant / day from shooting till 15 days prior to harvest.

Weeding

In the first four months after planting, it is necessary to remove weeds. The stirring of the soil is done by the spade after every 6 or 7 irrigations. It is necessary to maintain its tilth and absorptive power. Weeds can also be checked by the use of herbicides. Apply Diuron @ 4 kg per hectare and simazine @ 6 kg/ha. To control grasses and broad leaved weeds when applied after planting and repeated 30 days after planting. Glyphosate @ 1 kg a.i. per hectare at the time of planting followed by 0.5 kg a.i.per hectare at 30 and 60 days after planting of suckers is recommended. Banana is sensitive to 2, 4-D and hence the phenoxy compounds are not used.

Intercropping

Leguminous vegetables, beet root, elephant foot yam and sunhemp. Avoid growing cucurbitaceous vegetables.

Fertigation

Application of fertilizers through irrigation water is known as fertigation.

Advantages

- Very high yields with high quality fruits
- The dosages required by the plant can also be reduced by 25-50 % when the fertigation is practiced along with high density planting
- Effective utilization of fertilizers, no wastage
- Uniform growth of plants
- Reduced labour cost

It is possible to regulate and supply the required nutrients through irrigation water if drip systems have been laid out. This will help in obtaining very high yields compared to conventional method of fertilizer application in the soil. The dosages required by the plant can also be reduced by 25-50 % when the fertigation is practiced along with high density planting. Experiments conducted at this department showed that weekly scheduling of fertigation is sufficient and different nutrients.

25 litres of water / day + 200:30:300 g N:P_2O_5:K_2O /plant using water soluble fertilizer. For economizing the cost of fertilizers fertigate using normal fertilizers (urea and MOP) with 30% of the recommended dose along with recommended dose of P as basal at 2nd month of planting.

Fertigation schedule

Weeks after planting	N (%)	P_2O_5 (%)	K_2O (%)
9-18 (10 weeks)	30	100	20
19-30 (12 weeks)	50	-	40
31-42 (12 weeks)	20	-	32
43-45 (3 weeks)	-	-	8
Total	**100**	**100**	**100**

Flowers and Fruit

The inflorescence of banana (flowering stalk) emerges from the center of the pseudostem 7 to 9 months after planting; depending upon varieties by this time 26 to 32 leaves are produced. The process of banana flowering is called shooting. The flowers appear spirally along the axis of the inflorescence in groups of 10 to 20, covered by purplish-to-greenish fleshy bracts which shed as flowering development progresses. The first flowers to emerge are functionally female. In the edible cultivars, the rapidly growing ovaries develop parthenocarpically (without pollination) into clusters of fruits, called "hands." Although most banana cultivars produce seedless fruit, some are fertile and can set seed. The last flowers to emerge are functionally male. In plantains, the male part of the inflorescence and/ or male flowers may be absent or greatly reduced. The time from shooting to fruit harvest depends upon temperature, cultivar, soil moisture, and cultural practices and ranges from 80 to 180 days.

High Density Planting

To increase productivity in unit area, it is advisable to go for high density planting systems. Besides higher yield, high density planting also helps to reduce labour cost for matt management and increase the efficiency of utilization of inputs such as fertilizers and water. Experiments conducted by TNAU in cv. Nendran and Robusta with increased population per hectare has indicated that very high yields can be achieved. The population is increased by adjusting the row to row spacing and planting more than one sucker / hill. Accordingly, the following spacing is recommended. For cv. Nendran, adopt a spacing of 2 x 3 m with 3 suckers per pit (5000 plants/ha) with 25% extra NPK fertilizers per pit. In the case of Robusta, adopt a spacing of 1.8 x 3.6 m with 3 suckers per pit (4600 plants/ha) with 300:90:450 g of NPK per pit.

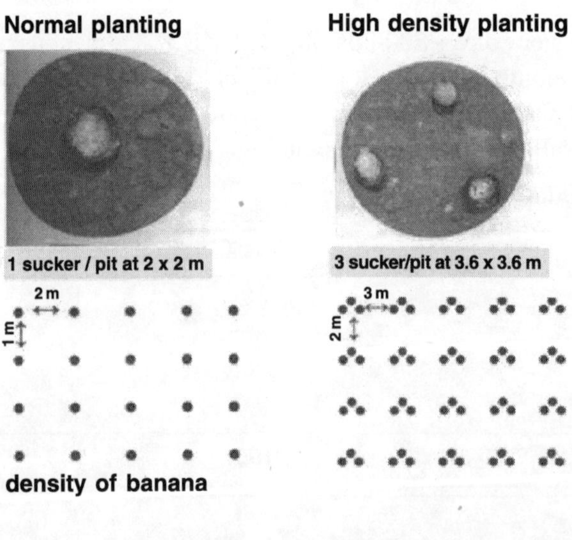

Fig. 6. Planting density of banana

Plants /Hill1.8 X 3.6 m (4800 pl/ha)

Paired row system 1.2 x 1.2 x 2.0m (5200 pl/ha)

Fig. 7. Different methods of HDP in Banana

SPECIAL INTERCULTURAL PRACTICES

Bunch Cover

During the development stages, the surface of the fingers may get bruised or damaged due to their exposure. Use of transparent polythene sleeves provided with 2-4% ventilation to cover the bunches can be helpful to avoid such damages. Besides, reducing the blemishes, it improves bunch appeal, advances the maturity by 7-10 days in banana. The right stage of bunch covering is when the last hand has opened and the male bud has to be removed at that time. The top end of the sleeves should alone be tied with the peduncle leaving the bottom open for better aeration.

Fig. 8. Banana Bunch cover

Removal of Male Bud

The removal of male bud along with withered styles 8-11 days after bunch emergence along with the removal of withered styles is recommended to favour fruit development. This is called denavelling

Removal of male bud

Male bud

Fig. 9. Denavelling in banana

Removal of Withered Styles and Perianth

In some varieties the styles are persistent and hence they remain in the bunch till maturity. They can be easily removed by a light brushing movement of the hand a few days after flowering and if it is delayed, it is difficult to remove and later, they become brown and shriveled. This operation eliminates the removal of infection by saprophyte fungi especially to avoid the fingertip disease.

Fig. 10. Dried and withered style and perianth

Desuckering

It refers to the removal of unwanted suckers. They are to be periodically removed otherwise they compete with the mother plant for nutrients, resulting in lower bunch weight and yield. Desuckering is accomplished by cutting the suckers at ground level or pushing a chisel-shaped crowbar between the parent and the sucker and giving a twisting motion. Under garden land and wetland conditions, desuckering is done at monthly intervals. In hill banana, two bearing plants and two followers per clump are considered optimum and the rest are desuckered. Normally, desuckering is done at 3 times in a year.

De-suckering

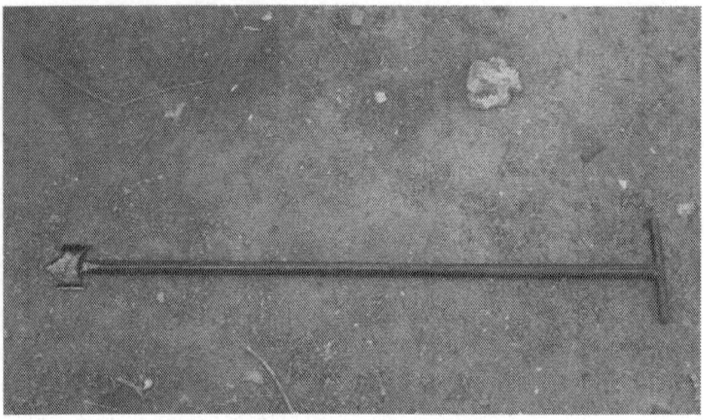

De-suckering equipment

Fig. 11. Desuckering in banana

Use Growth regulators

In certain pockets of Trichirapalli, the banana cultivar Poovan is with a peculiar development disorder *viz*., 'kottaivazhai' which is characterized by the presence of distinctly conical and ill filled fruits with a prominent central core having many under developed non-visible seed like structures rendering the fruits inedible. This disorder can be overdome by spraying 2,4-D 20 ppm when the last hand of the bunch is opened. The same chemical at the same dose and stage results in increased bunch weight and uniform grade especially cv. like Nendran and Monthan. b. In cv. Rasthali, hard mass or hard lump, characterized by pinkish brown, firm pulp than the usual soft pulp occurs, tests like immature or unripe fruits. Spraying the bunches uniformly with 2,4-D 1000 ppm or dipping the cut end of peduncle of the bunches for a period of 5 minutes appears to favour the reduction of lumps and improve the size.

The process of flowering governed by Gibberlin like substances helps in development of plant, later on anthecin hormone inducing flowering of plant, both combiningly called as "Dual factors hypothesis".

To improve the grade, 2, 4-D @ 25 ppm (25 mg/lit) – spray in Poovan & Co.1 after opening of last hand. It is poured in the growing apex, than bunch will have more of female flowers/fingers. This also helps to remove the seediness in poovan variety. 2, 4, 5-T – 100 ppm – increased finger size and appearance. 2, 4-D – 100 ppm – finger weight, pulp % and TSS. NAA 100 ppm – 5th and 7th month – increased fruit size and yield. Application of GA_3 at 50mg/L resulted in maximum yield and required less number of days for fruit maturity in Giant Governor Banana.

GA_3 400 ppm – 35 or 55 days – 3 times on alternate days - increased colour of fingers. GA_3 – 50 ppm – increased yield – 56-65 t/ha. Spray CCC 1000 ppm of 4th and 6th month after planting. Spray plantozyme @ 2ml/lit at 6th and 8th month after planting to get higher yield.

Sunscald

The peduncle of bunches may be covered with flag leaf to prevent 'main stock rot' and also the bunches with banana leaves to avoid sunscald.

Propping

The pseudostem requires the support from propping with bamboo or casuarinas poles at the time of bunch emergence. The pole is fixed and tied around on, one side of the plant or two poles are tied at the tip in such a way, they form 'x' shaped and are kept against the pseudostem to bear the weight of the bunches.

Fig. 12. Propping in banana

Mattacking

After harvesting, the pseudostem should be cut leaving a stump of about 60 cm height. This practice is called 'mattacking'. The food material stored in the left out stump continues to nourish the daughter sucker till it withers and dries up. Bananas are not usually allowed to ripe on the tree as it takes long time. Besides, the fruit peel splits, fruits ripens unevenly and fails to develop good colour and aroma. Hence, banana is made to ripen artificially.

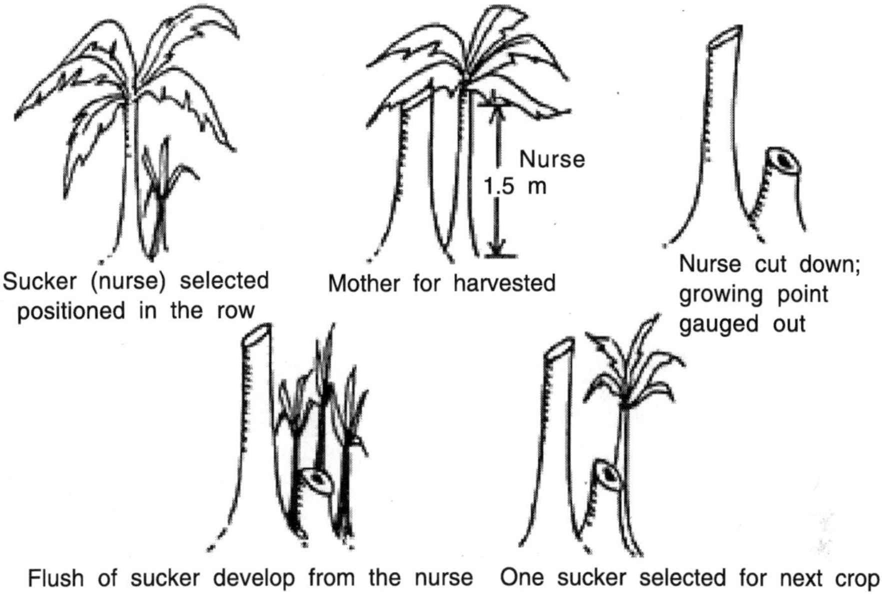

Fig. 13. Mattacking operation in banana

Smoking

Smoke treatment is the commonest method to induce ripening in Tamil Nadu and Maharashtra. Smoking is done with straw, leaves and cowdung in a closed chamber with bunches arranged in a heap for 18-24 hours in summer and 48 hours in winter. After taking the bunches from the chambers, they are placed in a well ventilated room for development of colour. Bunches take 3 days for ripening in this method. Ethrel treatment (2500 ppm) also cause ripening within 2 days. In this method, the bunches are arranged in a closed chamber and ethrel is taken in a beaker and added with few pellets of sodium or potassium hydroxide. This facilitates the release of 'ethylene' gas which aids in ripening the fruits.

HARVESTING AND YIELD

Maturity Indices

The colour of the fruit changes from deep green to a lighter green. The fruits are harvested when top leaves start drying. Shedding of floral ends of the fruits with slightest touch of the hand. The angles or ridges of the fruits become less prominent or they become round. Harvest may be delayed up to 100-110 days after opening of the first hand. Fruits are harvested when about 75 % mature as fingers on upper hands turn light green in color. The bunch is cut from pseudostem leaving a portion of peduncle for handling during transport.

Yield

The average yield is as follows,

1. Dwarf varieties : 30-40 t/ha/year.
2. Tall varieties : 15-20 t/ha/year.
3. Poovan, Rasthali and Ney Poovan: 40-50 t/ha/year.
4. Robusta and Dwarf Cavendish : 50-60 t/ha/year.
5. Monthan : 30-40 t/ha/year.

Post-Harvest Technology

Bunches should be kept out of light after harvest since this hastens ripening and softening. For local consumption hands are often left on stalks and sold. For export, hands are cut into units of 4-16 fingers, graded for both length and width and carefully placed in polylined 12 to 18 kg boxes. Prior to packaging, fruits are cleaned in water or diluted sodium hypochlorite solution to remove the latex and treated with thiobendazloe and other fungicides. It is successfully demonstrated that 100 gauge thick polythene bags with 0.2% holes enhance shelf life under room temperature as well as in cold storage.

Ripening

Bananas are not usually allowed to ripe on the tree as it takes long time. Besides, the fruit peel splits; fruits ripen unevenly and fail to develop good colour and aroma. Hence, banana is made to ripen artificially. Smoking done with straw leaves & cow dung in a closed chamber for 18-24 hours in summer and 48 hours in winter and later shifted to ventilated room for uniform ripening. The exogenous application of 100 ppm ethylene gas in an enclosed chamber for 24 hrs for will produce uniform colour and ripening. Fruits are ripened by exposure to ethylene gas (1000 ppm for 24 hrs) at their destination, in sealed "banana ripening rooms". Hands can be stored for a short time at temperatures not lower than 12°C since fruits are susceptible to chilling injury.

Storage

Banana can be stored at about 13°C with the Relative Humidity of 85-95 per cent for 3 weeks and is ripened in a week at 16.5 - 21°C. The fruits should not be stored / shifted under refrigerated condition. The storage life can be increased by keeping the fruits in high concentration of carbon dioxide and low concentration of oxygen. Also storing in sealed polythene bags containing ethylene absorbent like potassium permanganate. Shrink film wrapping or Waxol (12 per cent) treatment can extend shelf life up to 3 weeks.

PHYSIOLOGICAL DISORDERS

1. **Hard lump:** It is characterized by pinkish brown, firm pulp than the usual soft pulp occurs in cv.Rasthali, tastes like immature or unripe fruits. Spraying the bunches uniformly with 2, 4 D at 1000 ppm or dipping the cut end of peduncle of the bunches for a period of 5 minutes appears to favour the reduction of lumps and improve the size.
2. **Kotta vazhai:** It is also a malady of unknown etiology affecting Poovan banana. "Kottai" means seed, referring to conspicuously enlarged ovules and immature dark green fruits. Application of 2, 4-D 20 ppm is found to improve the bunch weight.
3. **Neer Vazhai:** It affects the variety Nendran. Infested plants show poor plant growth, delayed shooting, lanky bunch with few hands and immature unfilled fingers. Fruits ooze out watery fluid when cut and severe root damage is noticed. Cause of this disorder is unknown and found to be transmitted through suckers. Application of NAA improves the finger filling.

Package of Practices for Tissue Culture Banana

Advantages of using tissue cultured banana plants

- Plants are selected from high yielding clones so, yield will be high
- Plants are micro propagated under optimal conditions - Uniform superior planting material.
- Production of high yielding quality planting material which are uniformity healthy, disease free and having accelerated growth reducing the crop cycle.
- Stable plants with well-developed root system and their observation and growing capacity in very high.
- Tissue culture plants enable flexibility in accordance with the planting season and marketing demand.
- Fruit bunches are dense, and compact having well shape fingers for better marketability.
- Optimal yields are ensured every season following requisite cultural practices.
- Integration of 3 crops (including two ratoons) from tissue culture with advanced growth ensures highly profitable plantation.
- Large quantities of plants could be supplied at a time.
- Quality plant command premium in the market.

Propagation Using Tissue Culture Plants

At present, tissue culture plants of choice varieties of banana are commercially available. The cost of these planting materials are higher compared to suckers, but so also the benefits. The plants are initially free from disease inoculum and hence proved to be very vigorous once they establish in the field. If they are derived from superior and high yielding mother plants, then they also perform better yield wise. The plants usually attain shooting and harvest uniformly.

Compared to conventional suckers, the tissue culture plants will easily suffer post planting diseases due to soil borne pathogens such as Erwinia bacterial wilt. Hence, periodical drenching with bleaching powder @ 2 g/lit. at an interval of 10-15 days can be taken up if such infection is noted.

Varieties and Spacing

Sl.No.	Varieties	Spacing	No. of plants per acre	No. of plants per Ha.
1.	Dwarf Cavendish Basrai	(1.65m x 1.65m)	1440	3600
2.	Robusta, Poovan	(1.80m x 1.80m)	1210	3025

Land Preparation

- Plough and dig pits of 2' X 2' X 1.5' and fill the pits with well decomposed FYM/compost (15 -20Kg) and thoroughly mix with sand and soil in 1:1:1 proportion. Irrigate the pits to allow the soil to settle.
- Apply 1/2 Kg. neem cake and mix well with soil or apply Furadan 3 G @ 20g/pit or thimmet 10 G @ 12g/pit within a week after planting.
- We can also apply 20 g in each of *Azospirillum*, *Phosphobacteria* and 50 g VAM with 25 g of *Pseudomonas fluorescence* and *Trichoderma viridie* / plant at planting to control diseases and nematode.

Planting

Split open the polybag and remove the plant carefully with the ball of earth intact. Place the plant in the centre of the pit without disturbing the roots. Deep planting should be avoided. After a week of planting, drench the soil with Emisan @ 1g/litre. A week later spray streptomycin (500ppm) - 1/2 g/lr to control bacterial infection.

Instead of using these chemicals we can apply 50 g VAM in soil application with Application of *Pseudomonas fluorescens* and *Trichoderma viridie at* the rate of 5 g/ lit. as drenching (250 ml/plant) around the root zone may also be helpful.

Fertilizer Application

Fertilizer Application depends on soil fertility. General recommendation is 200:100:200 g/plant of NPK in split doses as detailed below:

S.No	Day's after Planting	Urea (g/plant)	Fertilizer Dosage(g/plant)		
			Ammonium Sulphate	Super Phosphate	Muriate of Potash
1.	30	45	100	125	50
2.	75	90	195	125	85
3.	110	110	245	125	85
4.	150	110	245	125	85
5.	180	90	195	125	100
6.	At Bunch emergence	–	–	–	85

Apply fertilizers 2" to 3" below the soil surface. Apply 1st and 2nd dose of fertilizer at a distance of 20 cm away from the plant. 3rd and 4th dose of fertilizer should be applied at 40-50 cm and 5th and 6th dose at 50-60 cm. from the plant.

❑❑❑

CHAPTER 9

CITRUS

Botanical name: *Citrus spp.*,

Family: Rutaceae

Chromosome number: 2n-18

Origin: South East Asia and Indio-china

Citrus fruit is the world's leading tree fruit crop and in India, it is the third important fruit crop next to mango and banana. This group of fruits includes sweet organs, mandarins, limes, lemons, grapefruits, pummelo, citro Kumquat etc. It is believed that most of the species under the genus *Citrus* are native to tropical and subtropical regions of South – East Asia, particularly India, China and in the regions between these two countries. Now citrus is grown in more than 100 countries of the world and the major citrus producing countries are USA, Spain, India, Italy, Israel etc. Citrus fruits contain considerable amounts of Vitamin C. Fruits are also good source of Vitamins and P. The mild bitterness in juice is due to the presence of glucoside called Naringin which is said to have a medicinal value. The rind of the citrus fruits is rich in pectin and essential oils. The major mandarin growing area in India are Coorg (Karnataka state), Nagpur belt (Maharashtra), Darjeeling District (West Bengal) etc. In Tamil Nadu, mandarins are grown in Pulney, Nilgiris and Shervaroy hills.

Citrus plants are perennial, small trees. Its leaf is uniformly compounded. In most species, the petioles are winged exception being lemon and citron. Grapefruit and pummelo have large wings while sweet oranges and acid lime have small wings. Flowers are produced on current season growth in cymes both axillary and terminally. Two types of flowers, perfect and imperfect (with rudimentary pistils) are found. The fruits of citrus are of a special type of berry known as 'hesperidium'. The pericarp is divided into three layers.

a. **Exocarp (Flavedo)** - Outer most layer of the fruit having many oil glands.

b. **Mesocarp or albedo -** is a white spongy portion adhering to the inner side of the exocarp.

c. **Endocarp -** is the edible portion containing many carpel segments. The segments are separated by carpel walls. Within the carpels are many juicy vesicles developed from hair – like papillae on the segment membrane and some seeds. The juicy vesicles are attached to the membrane with the thread like stalk.

Nutritive value

They are a rich source of vitamin-C and protect a person from scurvy a dreaded disease causing body sores, bleeding gums, etc. The citrus fruits also contain vitamin K, which keeps the small blood vessels in our body in healthy condition and helps in assimilation of vitamin C. The oranges of the reticulates group are good sources of beta-carotene or vitamin A. Citrus fruits also have fair amounts of most mineral nutrients like calcium. Perhaps that is why citrus is important in diets of sick and convalescents. All the citrus fruits supply a large amount of potassium and calcium.

The important sugars in citrus fruits are sucrose, g1ucose and fructose occurring in a ratio of 2:1:1 respectively. Thus, they are good energy food too. The acidity of these fruits is mainly due to citric acid. In the sweet varieties, acidity comes down upon ripening and at the same time fruit's sugars increase. Most citrus fruits have trace of bitterness which is due to the presence of flavonoid compounds. The aromatic principle in these fruits is due to a chemical substance called limonene and other essential oils. All these chemical compounds present in citrus fruits are found to have extensive uses in industry, pharmaceuticals and food processing.

Sub Groups in Citrus

1. **Acid group**
   a. Acid lime: *Citrus aurantifolia*
   b. Tahiti or Persean lime: *Citrus latifolia*
   c. Rangpur lime: *C. limonia*
   d. Lemon: *Citrus limon*
   e. Rough lemon: *C. jambhiri*
   f. Citron: *C. medica* (Kidarankai in Tamil, used for pickling)
   g. Sweet lime: *Citrus limettoides*

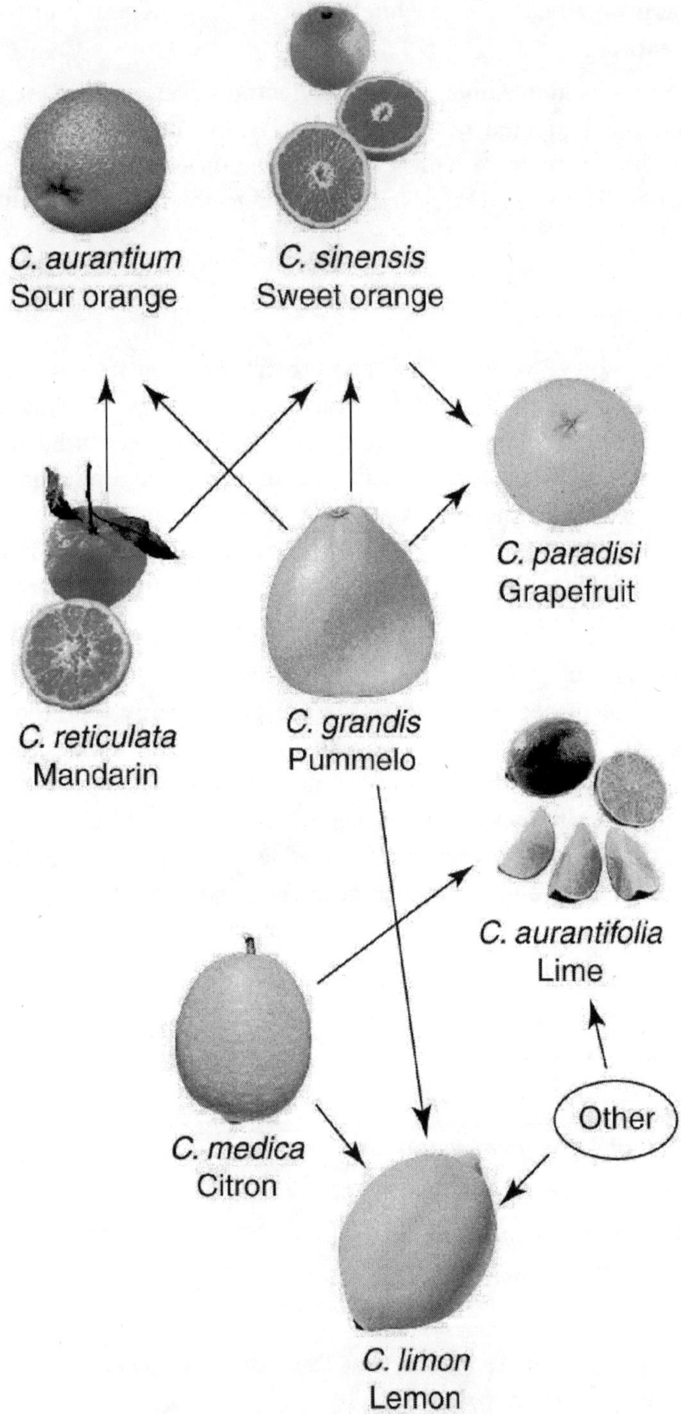

Fig. 14. Sub Groups in Citrus

2. **Orange group**
   a. Sweet orange: *Citrus sinensis*
   b. *Sour orange: Citrus* aurantium (Narthankaai in Tamil, used for pickling)
   c. Multiple leaf orange: *C. multifolia*
   d. Japanese summer grape fruit: *C. natsudaidai*
3. **Mandarin group: (Loose jacket)**
   a. Coorg mandarin, Nagpur Santra and Kodai orange: *C. reticulata*
   b. Japanese Satsuma mandarin : *C. unshiu*
   c. Willow leaf mandarin: *C.deliciosa*
   d. King mandarin: *C. nobilis*
   e. Kinnow mandarin: King x willow leaf
   f. Tangerine orange var Dancy (trifoliate x mandarins): *Citrus tangerina*
4. **Pummelo and grape fruit group**
   a. Pummelo: *C. grandis*
   b. Grape fruit: *C. paradisi*
   c. Kumquat: *Fortunella sp.*
5. **The fifth group consists of mainly hybrids of different citrus fruits with trifoliate orange** (*Poncirus trifoliata*) and mainly used as rootstock.

   e.g. Citrange (*Poncirus trifoliata x C. sinensis*) var. Troyer, var. Carrizo

   a. Citrangor (Citrange x *C. sinensis*)

   b. Tangelo (Tangerine x grape fruit)

   c. Citrangequat (Citrange x kumquat)

Major Citrus Species of Horticultural Importance are

a. **Acid Lime (*C. aurantifolia*):** A highly polembroyonic species, flowers are white, fruits are small, round to oval, greenish yellow and thin skinned, core, solid at maturity e.g. kagzilime, PKM-1 developed at HCRI, Periyakulam, a selection from seedling progenies of Kadayam type, fruits are big, each weighs about 50 g, rich in juice (52%). Each tree bears on an average 934 fruits weighing 37 kg.

b. **Sweet Orange (*C. sinensis*):** Also known as tight skinned oranges. A highly polyembryonic species, trees medium – large, fruits sub-globose to oval in shape, orange colour skinned, segments 10 -12, orange coloured, pulp juicy. Important cultivars are Mosambi, Malta Blood Red, Sathugudi, Pineapple and Valencia Late etc.

Acid lime

Sweet oranges

Mandarin

Pumelo

Grapefruit

Fig. 15. Different groups of citrus

c. **Mandarin Orange (*C. reticulata*):** Also known as 'Tangerine' or loose skinned oranges. Fruits medium sized, globose, sweet in taste, segments easily separable, core open at maturity, loose skinned orange in colour, rind thin, rind and segments easily separable, usually 10 – 14 segments in each fruit. Important cultivars are Coorg Orange, Kodai Orange, Nagpur Santra, Khasi Mandarin etc.

d. **Lemon (*C. limon*):** The petals are pink coloured and the fruits are oval to elliptic with prominent nipple. Fruit's surface smooth, light yellow, core solid and juice abundant and acidic. Important varieties are Sevelle, Malta, Rajamundhry, Mayor, Eureka etc.

e. **Pummelo (*C. grandis*)**: A monoembryonic species with large sized fruits, sub- globose to pyriform in shape, with thick and spongy rind, white fleshed and pink or red fleshed types are available.

f. **Grapefruit (*C. paradisi*)**: A polyembroyonic species, fruits large, sub- globose in shape, fruit rind and flesh are yellowish in colour, fruits highly juicy, sweet with bitter taste, central axis or core opens at full maturity.

Climate and Soil

Citrus can grow well in a wide range of soil but it prefers deep, well-drained loamy soils. The best pH is 5.5 to 7.5 good drains is an important pre-requisite of soils and the water table should be always below 2.0 meters.

Citrus thrives well in frost - free subtropical to semi-tropical climate. However each citrus kind needs a specific climate for its better performance. Sweet oranges prefer dry arid conditions coupled with well-defined summer and winter seasons having low rainfall ie 20-40 cm. Places receiving high rainfall and high humidity are unsuitable. On the other hand mandarins perform well in sub-mountain tracts like Coorg, Wynad, Nilgiris, Pulney and Shervaroys hills ranging from 600-1000 m and the rainfall ranging from 75 to 250 cm. The do equally perform well in Nagpur area with an elevation of 300-600 m, the rainfall of 110 cm, relative humidity of 53 per cent besides extremity in temperature ie maximum temperature being 47°C and the minimum temperature being 6°C. Lime and lemon on the other hand prefer warm regions, moderately moist and free from strong winds. Punmmelo and grapefruit can thrive both under humid and dry tracts. They can grow well upto an altitude of 1000 m.

ACID LIME

Botanical name: *Citrus aurantifolia swingle*

Chrosome No.: 2n = 18

Origin: India

Acid lime also known as Sour lime, *Kagzi* lime or Mexicain lime Keylime, West Indian lime etc. In Hindi it is calles as *Neebu.* It is a good source of vitamin - C and extensively used for culinary purposes. Salted lime peel is recommended for indigestion. It is also have medicinal valu, it is an appetizer. India is the largest producer of acid limes in the world, but ranks fifth in the production of limes and lemons. Acid lime is the third important citrus fruit after orange and mandarins. It is grown in Andhra Pradesh, Gujarat, Tamil Nadu, Karnataka, Maharashtra, Assam and Rajasthan.

Climatic and Soil Requirement

Acid lime is tropical in its climatic requirement and sensitive to frost than any other citrus fruits. It thrives well in a dry climate. The optimum temperature is 20-30°C. Right from sea level upto even 1500 m above MSL it can be grown. Lime grows well in deep (2-2.5m) well drained soils rich in organic matter. pH range of 6.5 to 7.0 will be ideal for better growth and yield. It can not grow in water logged soils, alkaline soils with high lime content. Lemons grow better in shallow soils also if well drained.

Important Vatieties

i. **Kagzi lime** is the best most popular one.

ii. Some of the improved varieties of limes are: **Pramalini, Vikram and Saisarbati** - Supposed to be canker free and prolific bearers.

iii. **Chakradhar:** Seedless strain of acid lime,

iv. **PKM-1** - Heavy yielder.

v. **Selection 49:** Tolerant to canker, tristiza and leaf miner. It is also a prolific bearer.

vi. **Seedless lime:** Fruits are oblong and prolific bearer.

vii. **Tahiti lime** - Plants nearly thornless, leaves much larger of different shapes. Fruits also much larger and seedless - Triploid.

viii. **Mithachikna (Thin rind) and Mithotra (Thick rind):** Two varieties of sweet lime with less acid, sweet fruits and are mostly used as stocks.

ix. **Rangapur lime** is grown for root stock purpose.

Varietal Description

i. **Kagzi lime:** It is the most common variety found in north India. But this cultivar is highly susceptible to tristeza virus and bacterial canker diseases.

ii. **PKM - 1:** It was developed at Horticultural College and Research Institute, Periyakulam of Tamil Nadu Agricultural University by selection from seedling progenies of kadayam type of Nellaikattabomman (Thirunelveli) district of Tamil Nadu. The fruits are rich in juice (52.3% by weight). Acidity of juice is 6.5%. Each tree gives on an average 934 fruits per year weighing 36.9kg.

iii. **Sai-Sharbati:** It is a kagzi lime selection developed at MPKVP, Rahuri, Maharastra from a local germplasm collected from Western Maharastra. Fruit surface smooth, fruits more uniform good size, thin skin, high juice, TSS and acidity. High yield potential with higher (25%) cropping intensity in summer. Tolerant to canker and tristesia.

iv. **Tahiti Lime or Persian Lime or Seedless lime:** It is closely related to lime but belongs to *Citrus latifolia.* It is triploid and does not possess seeds, hence, multiplied by ground layering and air layering.

v. **Rangpur Lime:** It is also related to acid lime but belongs to *Citrus limonia.* It is mainly used as rootstock as it has good adaptability to a wide range of soil conditions especially for heavy soil. It is tolerant to tristeza virus and also to salt. However, it is susceptible to foot rot, exocorits and xyloporosis.

PROPAGATION

Raising of Seedlings

It is commonly propagated by seed all over the country because the seeds exhibit a high percentage of polyembryony. Acid lime is mainly propagated through seeds selecting the nucellar seedlings. Treat the seeds with GA_3 25 ppm (or) thiourea 1 % - improves germination. Seeds should be sown on raised nursery bed immediately after extraction of seed (at least within 3 days) as they lose their viability quickly. Since, it produces polyembryonic seedlings which are from nucellar origin. Each polyembryonic seed produces 3 to 4 seedlings. Among them, one is sexual origin remaining are nucellar. Nucellar or apogamic seedlings breed true to type not in the case of sexual seedlings. The sexual seedlings are stunted and poor in growth compared to apogamic or nucellar seedlings. When the seedlings attain 4-6 leaf stages, they are transplanted in secondary nursery at spacing of 20-30cm and off type or sexual (hybrid) seedlings are identified and removed carefully. Seedlings are ready for transplanting 6 to 9 month after sowing. However, usually one year old seedlings are preferred for transplanting, Acid lime can also be propagated by budding, ground and air layering. Tahiti lime is seedless and hence propagated by air layering. Acid limes can also be budded by patch budding on Rangpur lime (*Citrus limonica*), Rough lemon and Trifoliate orange (*Ponicirus trifoliate*).

Cross Protection or Pre-Immunization

As the plant are susceptible to a viral disease viz., 'tristeza', pre-immunization by inoculating a mild strain of this virus through patch budding of a bark piece on healthy seedlings is done to induce resistance to virulant form of virus. Such seedlings are called Pre Immunized Acid Lime Seedlings (PIALS)

Field Preparation and Planting

Field is ploughed 2-3 times and pits of 75 cubic centimeters are dug at a spacing of 3.6 m x 3.6 m or 4.5 m and 5 m x 5 m depending on the fertility and type of soil. A close spacing of 3 m x 3 m can be adopted initially and after about 8 years alternate rows can be removed. This will increase the yield in early years.

Planting

After filling the pits with top soil and 20 kg of FYM, the seedlings (10-12 months old) can be planted in the centre of the pit and supported by staking with bamboo stick to prevent any possible break by wind. The pits are irrigated immediately after planting.

Season of Planting

Planting is done during monsoon season. September – October or February – March, when assured irrigation facilities are available.

Manures and Manuring

A bearing lime tree is to be fertilized with 600g of N, 300g of P and 300g of K besides 50kg of Farm Yard Manure. Apply NPK along with VAM (*Glomus fasiculatus*) @ 1 kg/tree will help to accumulate Phosphorus, Zn, Cu and sulphur.

The total nutrients have to be split into two equal parts and applied once during June -July and other during November - December. The manures and fertilizers are mixed and applied as a band application in the basin 70 cm away from the trunk and incorporated. NPK @ 600:300:300g + 15kg neem cake / plant / year significantly increased the fruit yield and quality in acid lime. Application of green leaves 30 kg per tree once in 3 months also recommended.

Manures and Fertilizers	1 year (kg)	Annual increase (kg)	From 6th year (kg)
FYM	10.00	5.00	30.00
N	0.200	0.100	0.600
P	0.100	0.025	0.200
K	0.100	0.040	0.300

Micronutrient Application

The spray solution containing following micronutrients can be applied once in three months at the time of new flesh production. Zinc sulphate - 0.5%, Manganese - 0.05%, Iron - 0.25%, Magnesium - 0.5%, Boron - 0.1%, Molybdenum - 0.003%. In addition to that apply 50 g in each of ZnSo4 and Fe So4 per tree per year.

Irrigation

Under tropical condition water requirement of acid lime has been found to be 875 mm per year. The daily water requirement would be 50-60 litres per plant during humid months and 90-100 litres during summer. The chloride content of the irrigation water should not exceed 100 ppm. Soil moisture stress will be detrimental to plant fruit enlargement and juice formation.

Interculture and Weeding

Under normal planting system, pulses like black gram, cowpea or vegetables like cucurbits, (beans and carrots during winter season) can be grown as intercrops. Mulching basins will be very beneficial. Weed management can be done by a pre-emergent application of 3 kg diuron per ha followed by gramaxone 1.5 lit as a post emergent application in the interspaces.

SPECIAL HORTICULTURAL PRACTICES

Crop Regulation

Paclobutrazol @ 18g/ tree as soil application enhanced the flowering, fruit set and produced 1600 fruits / tree / year in acid lime. Flowering can be regulated by with holding irrigation for 1-2 months and again stress is alleviated which will result in flowering. But excessive stress will lead to over flowering, poor fruit set and plant health will also be affected adversely. Application of 500 ppm of paclobutrazol (PP^{333}) at new flush will encourage profuse flowering. Application of 2, 4, 5-T

@ 20 ppm at pepper stage would help to prevent fruit drop. To increase fruit set spraying 2,4-D-20 ppm during flowering. Fruit retention spraying- 2,4-D@ 20 ppm or NAA 30 ppm after fruit set (marble size).

Growth Regulators

GA_3 - good for staggering fruit production. Cultar @ 50 ppm at new flush good flower induction. 2,4-D @ 20 ppm at flowering – fruit set increase and fruit drop reduced. NAA 30 ppm – reduces fruit drop. Spray 2,4 -D @ 20 ppm during flowering to increase fruit size. Spray NAA 30 ppm or 2, 4 – D 20ppm at fruit set (Marble size) for fruit retention.

Pruning

Remove branches of main stem upto 45 cm from ground level and remove the water sprouts, dead and diseased branches.

HARVESTING AND POST-HARVEST TECHNOLOGY (PHT)

Maturity Indices

Unlike some other fruits, citrus fruits do not ripen further once they have been removed from the tree, so it is important that they are picked at the right stage of maturity. Maturity is measured depending on different characteristics such as color, juice content, level of soluble solid (sugar) and solids to acid ratio. Normally, citrus fruits are harvested by hand. Though limes can be harvested throughout the year, the main crop is harvested during different periods in different parts of the country. The average yield is 20-25 kg/tree/year. Harvesting season extends from Jan-September depending upon the locality for acid lime and May for lemons, June-Aug for Rangpur limes. Lime yields 2000-5000 fruits/plant.

Storage

Acid limes can be stored for 6-8 weeks at 8.3 to 10°C and 85.9 % RH and lemons for 8-12 weeks at 7.2-8.6°C and 85-90 % RH. Limes can be stored at 18°C. At Horticultural College and Research Institute, Periyakulam a low cost storage tank has been developed with double layer brick work, the interspace filled with sand which is kept wet by periodical watering. This is called as 'Zero energy cool chamber'.

SWEET ORANGE

Botanical name: *Citrus sinensis*

Chromosome No.: 2n = 18

Origin: Asia

Sweet orange is also called as tight skinned orange or tight jacket orange. Due to sweet taste, pleasant flavour and refreshing qualities, the demand for orange as a dessert fruit is very great. It is a good source of vitamin C (64 mg / 100 g of juice) besides vitamin A and B. Major Producing States - Andhra Pradesh, Karnataka, Maharashtra, Punjab and Rajasthan.

Climatic and Soil Requirements

Sweet orange thrives better in a subtropical dry climate (low humidity) having well defined summer and winter seasons, coupled with low rainfall. Rainfall seems to be unimportant if irrigation is provided. Even in a tropical climate it can be grown. But atmospheric humidity exerts a greater influence. Though it can be grown in a wide temperature range of 10ºC to 38ºC, the optimum temperature would be 20 – 30ºC. Sweet oranges do not bear fruits successfully in regions of heavy rainfall and high atmospheric humidity like Konkan, Kerala, Mysore, East Coast of Tamil Nadu, West Bengal, Bihar and Assam. Here trees flower irregularly at different times and fruits do not develop colour and ripen properly. Dry climate of Pune, Ahmedabad and Hyderabad (RF 500 mm to 1750 mm / annum restricted to June - September) favours the cultivation of an excellent variety Mosambi. In Cuddapah, Kurnool, Chithoor and Rayalaseema of Andhra Pradesh, a variety "Sathgudi" is extensively cultivated. "Malta Blood Red" is a popular variety grown in plains of Punjab, Rajasthan and Uttar Pradesh where summer and winter are marked and severe. Ideal soil for sweet orange cultivation would be minimum or light loam. The deep alluvial loam of Indo - Gangetic plains and coarse friable black soils of Madhya Pradesh are considered excellent for sweet orange. It is very sensitive to high concentration of salt and cannot withstand water logged condition. The best pH would be 4.5 to 6.5.

Varieties

Blood red in Haryana, Punjab and Rajasthan. Jaffa, Hamlin, Pineapple are exotic, Mosambi in Maharasthra and Sathgudi in A.P.

Sathgudi

It is the most important sweet orange cultivar grown in Tamil Nadu and Andhra Pradesh. A good variety with fruits having a fairly smooth surface which is green at full maturity and turn to pale orange at sesnescence. Flesh is pale yellow in colour and sweet in taste with characteristic flavour. Granulation viz., drying of juice vesicles is a problem in this variety.

Mosambi

It is the commercially grown excellent early season variety produced under irrigation in the dry climate of Pune, Ahmedabad and Hyderabad. Fruits are small to medium and sub globose having smooth surface with longitudinal furrows. Apex is marked with circular ring. Flesh is pale yellow, juicy, acidic and has no marked flavour. Being an early variety reaching the market in November itself,it fetches a premium price. The TSS / acid ratio is 13 :1.

Pine Apple

It is a midseason variety, the fruits reaching the market in December. Fruits are medium sized, having a distinct flavour and deep golden orange colour. The shape is round to ovate. Thin rind is tight and bright. Apex is rounded or slightly depressed. Flesh is orange yellow in colour; juice has a TSS / acid ratio of 14 :1

Jaffa

It is another midseason variety (December maturing) grown in Punjab, Northern Rajasthan and Western Uttar Pradesh. The fruits are round, orange yellow to orange red with smooth skin. Flesh is light orange in colour; juice has a TSS / acid ratio of 14 : 1.

Malta Blood Red

It is an excellent variety favoured for its red flesh. Extensively grown in the plains of Punjab, Rajasthan and Uttar Pradesh where summer and winter are more marked and severe. Fruits are medium to large, round to slightly oblong. Fruit skin is cadmium yellow to deep orange, glossy,thin and tight. Flesh is streaked with red colour in early ripening period. After full ripening, the flesh becomes completely red. TSS / acid ratio is 14 : 1. The maturity time is January. Though it is a shy bearer, because of good quality it is highly priced.

Valencia Late

It is one of the commercial varieties grown in USA. Fruit shape is round to slightly oblong having a deep golden orange thin skin. Juice has a TSS / acid ratio of 10 : 1. It is a late variety maturing during February - March.

Washington Navel

It is a seedless budsport of the Brazilian Orange variety "*Laranja S electa*". It is characterized by a rudimentary secondary fruit embedded in the fruit apex. It is commercially cultivated in USA.

Other Important Varieties are

Shamouti of Israel, *Succari* of Egypt and *Dobla Fina* of Spain.

Propagation, Preparatory Cultivation and Planting

Sweet orange varieties are propagated by patch budding or shield budding on selected root stock seedlings of one year old. The buds should be obtained from true to type, high yielding, disease and virus free mother tree which was indexed and certified as virus free tree. The most ideal time is September to November.

Root-stocks recommended

Sour orange: (*Citrus aurantium)*

It is widely used as root-stock for sweet orange. Trees on sour orange give high yield with quality fruits. But it is susceptible to tristeza virus, citrus nematode and burrowing nematode.

Rangpur lime: (*Citrus limonia)*

Tolerant to tristeza virus and salt. But susceptible to foot rot, exocortis and xyloporois.

Carrizo citrange

(*Citrus sinensis* 'Washington' sweet orange x Poncirus trifoliata) Citranges are recently recommended rootstocks in the place of sour orange. Citranges are tolerant to tristeza, intermediate resistant to root rot and resistant to burrowing nematodes.

Cleopatra mandarin: (*Citrus reticulata)*

Tolerant to tristeza, exocortis, xyloporosis and fairly tolerant to phytophthora root rot. As a rootstock, it excludes sodium and chloride from the soil.

Rough lemon or Jambhiri: (*Citrus jambhiri)*

Tolerant to tristeza, relatively tolerant to saline and calcarious soils but susceptible to root rot and blight.

Sweet orange: (*Citrus sinensis)*

On its own root-stock, the plants are cold hardy and have wide range of soil adaptation, resistant to tristeza, exocortis and xyloporosis. Recent studies in Mayal (Karnataka) indicated that sweet orange when budded on wood apple (*Feronia limonia)* resulted in dwarf (1 M height) and bushy (1 square metre spread) trees. They are resistant to drought and to soil borne diseases. Bearing and quality of fruits are also good. In the selection of bud from the scion of the mother tree, care should be taken to confirm true to type and free of virus. More than 15 viruses are reported in the world among which four namely tristeza including seedling yellow strain, greening (mycoplasma disease), exocortis and a strain of psorosis are found in India. Buds are taken from such trees which are indexed as free of viruses and budded on root stocks raised. Budded plants will be ready for planting in the main field in about 12 months period.

Planting

Budded plants can be planted in pits of 1 cubic metre size filled with 20 kg FYM and top soil dug at a spacing of 6M x 6M. A close spacing of 6 M x 3 M (as double planting) can be adopted. After 8 – 10 years, alternate plant or alternate row on either way can be removed. This will increase the yield in early years.

Nutrition, irrigation and interculture

One tonne of orange would remove 1. 18 - 1. 85 kg N, 0.17 - 0.24 kg P, 1.79 - 2. 61 kg K, 0.36 - 1.04 kg Ca and 0. 17 to 0.19 kg Mg. A quantity of 800 g of N, 200 g of P and 400 g of K / tree / year has to be applied in two equal splits one during June and another during December.Micronutrients like Zn, Mn, Mg @ 0.5 % foliar spray and Cu @ 0.25 % have to be applied once during March and second during July so as to avoid any micronutrient deficiency. Foliar spraying should be done after new flushing (on 30 - 45 days old leaves). Irrigation must be done when the available soil moisture (ASM) at surface soil level (0 - 30 cm) reaches 50 %. In sweet orange cv. Sathgudi, castor cake @ 7.5 kg along with 400 : 150 : 300 g NPK per plant year improved the fruit yield and quality. To control fruit drop, 2, 4 - D @ 10 ppm (10 mg / litre) should be sprayed two times, once in the first week of September.

Irrigation

Irrigation requirements depend upon soil and weather conditions. Irrigations should be regular during fruit development. Water should never come in direct contact with the trunk of the tree; For this reason double ring or check bund method is best suited for this crop. Presently drip irrigation is becoming popular which helps in saving of irrigation water. Stopping irrigation 1 or 2 months prior to flowering is beneficial to the crop, till the tree withers and drops half of its leaves.

Interculture

During the pre-bearing stage of the plants, vegetables other than solanaceous crops can be grown, taking care not to waterlog the soil around the trunks of the plants. Leguminous crops are the best Intercropping with „pea was found to improve the yield of sweet orange. Cucurbits also can be grown successfully.

Weed Management

Weeds can be controlled with pre-emergence spray of diuron @ 3Kg/ha twice at120 days intervals. Other weedicides used are simazine, atrazine, bromacil, 2, 4-D. etc.

Pruning and Training

No regular pruning except removing dead, diseased and over crowding branches after harvesting of the fruit. Plants should be trained during first 3 years to have a well distributed frame work at 1mt height on a single trunk.

Harvest, yield and Storage

For sweet oranges the main harvesting season in North India is from December to February while in the South, it extends from October to March. The fruits remain fresh for several weeks without any deterioration after attaining full maturity which facilitates harvesting according to one's choice and demand in the market. Application of 10 ppm of GA_3 by spraying on the trees just before fruits turn yellow will not only help to extend harvesting season but also improves resistance to fruit flies, transport tolerance and resistance to post harvest moulds and mechanical injury by keeping the peel tough. The yield is 200 - 300 fruits / year / tree. Oranges can be stored for 12 weeks at 5ºC (If stored at 0ºC, there will be chilling injury).

PHYSIOLOGICAL DISORDERS

Granulation

Granulation is a physiological disorder of the juice sacs of citrus fruits wherein they become comparatively hard, dry, assume a grey colour and become enlarged.

It is common in varieties like Sathgudi, Mosambi, Jaffa, Malta Blood Red and Valencia late. To decrease this disorder, irrigation (amount and frequency) should be reduced. The trees should not be over irrigated. Lime spray @ 20 kg in 450 liters of water controls this. Zinc (0.5 %) spray and copper (0.25 %) spray also control this.

Colour Development

Colour development in sweet orange especially under tropical condition is a problem and to obtain colour on fruit, 1000ppm of ethephon as a pre-harvest spray or fruit dip in combination with Benlate 1000ppm is recommended.

Sun Burn

Sun burn or sunscald is another problem in sweet orange fruits coming to harvest during April to June. The portion exposed to sun develops yellow patches which gradually turn brown and become hard. The inner portion becomes desiccated and discoloured. Affected fruits are malformed and have low juice content. Severely affected fruits drop off and leaves turn brown. This physiological disorder can be controlled by,

1. Spraying the trees with lime solution @ 20 g / litre before summer.
2. Planting a row of *Sesbania grandiflora* or *Sesbania aegyptiaca* on the South Western side of the tree in the basin at an espacement of 30 cm.
3. Regulation of irrigation to bring down the temperature.
4. Mulching the tree basins with groundnut husk or haulms.
5. Raising line wind break with teak plants to check hot dry winds.

MANDARIN ORANGE (LOOSE JACKET ORANGE)

Botanical name: *Citrus reticulata*

Family: Rutaceae

Chromosome no: 2n - 18

This group of orange is otherwise called Kamala orange, Nagpur santra of Maharastra, Coorg of Karnataka and Kodai orange of Tamil Nadu, Japanese satsuma, King, Willow leaf of USA, Kinnow of Punjab and Tangerine of Florida. Belong to this group though they are classified under different botanical names. This group is characterised by the loose skin of the fruit.

Climatic and Soil Requirements

Mandarin oranges grow very well under a mild subtropical condition preferably free from frost. They prefer more humid (as against more arid by sweet oranges) and subtropical summer but warm winter and would grow successfully in a sub-mountain tract (600 m to 1100 m above MSL) with an annual rainfall of 750 mm to 2500 mm. Such conditions are met with Coorg (Mysore), Wynad tract of Malabar (Karnataka), Lower Palani and Shervoroys of Tamil Nadu, West Bengal at altitudes below 300 m MSL. Severe frost during winter and heavy hot winds during summer will cause adverse effect on the growth of mandarins. A medium or light loam with a pH range of 5.5 to 6.5 would be ideal to grow mandarin orange.

VARIETIES

Kodai Orange

Trees are vigorous; fruits are very small characterized by loose rind and medium flavour. But it is a heavy seeded variety.

Nagpur Santra

Large vigorous growing variety. Fruits are medium to sub globose having a loose rind. Flesh is fine textured with abundant juice. Fruits mature in January – February.

Coorg Mandarin

Medium to large fruits, bright orange with a loose rind. It has a good flavour, ripens later than Nagpur Santra. It is a regular bearer. It is the commercial variety in the Coorg region of Karnataka. In Darjeeling district of West Bengal, the variety grown is known as Darjeeling orange and it is Desi in Punjab. In Sikkim, the mandarin cultivar grown is known as 'Sumithira' while in Meghalaya it is called as 'Khasi Mandarin' or 'Soh Niamtra'.

Satsuma Mandarin: (*Citrus unshiu*)

It is a Japanese variety with small spreading tree. Fruits are seedless with thin rind having orange colour at maturity. The quality of fruits is excellent with good blend of sugar and acidity.

King Mandarin: (*Citrus nobilis*)

It is a commercial variety of USA. The tree grows 5-6 m, petioles narrowly winged. Small, flattened, orange red fruits with distinctly sweet pulp, which is juicy, and of excellent quality.

Willow Leaf Mandarin: (*Citrus deliciosa*)

It is another commercial variety of USA. Trees medium sized with drooping growth habit. Fruits strongly compressed and yellow to light orange in colour with distinct flavour. The fruits are juicy, rind and segments loose and seedy.

Kinnow: (King x Willow Leaf)

It is a hybrid of King x Willow leaf. It has performed very well in Pakistan, Punjab and foothills of Himachal Pradesh. It has medium sized fruits with orange skin colour, flesh bright and golden yellow. Segments do not separate easily. TSS 10o brix, acidity 0.8%.

Dancy Tangerine: (C.tangerina)

This is the commercial variety of Florida in United States.

Propagation

Mandarin varieties are propagated by shield budding or 'T' budding using the buds on Rangpur lime (*Citrus limonia*) or Cleopatra Mandarin (*C. reticulata*). Bud wood taken from the disease free mother plants. While Rangpur lime is a vigorous, hardy rootstock with good adaptability to a wide range of soil particularly heavy soil, tolerant to tristeza and salt, it is susceptible to foot rot, exocortis, and xyloporosis. Cleopatra mandarin is the most salt tolerant rootstock with the ability to exclude sodium and chloride taken up by root system. It is also tolerant to tristeza, exocortis and xyloporosis and fairly tolerant to phytophthora root rot. For Kinnow (Hybrid mandarin), the best rootstock was Jatti Khatti (*C. jambhiri*) and Karna Khatta (*Citrus karma*). Rootstock seedlings are raised in nursery and after 3-4 weeks time they will germinate. When the seedlings attain 4-6 leaf stages, they are transplanted in secondary nursery at spacing of 20-30cm and off type or sexual (hybrid) seedlings are identified and removed, by which true to type nucellar seedling rootstocks are obtained. 'T' budding or patch budding is done on one and a half to two years old seedling. In about 6-9 months, the budded plants will be ready for transplanting in the main field. Troyer citrange rootstock is ideal rootstock for Kinnow mandarin.

Preparation of Main Field and Planting

The budded plants can be planted during June to December. Pits of 45 cubic centimeters are dug at a spacing of 6 m x 6 m and filed with farmyard manure, sand and top soil and then basins are formed. The buddlings are planted in the centre of the pits and irrigated.

Manuring

The following nutrients are applied in two splits one in May-June and another in September-October.

Fertigation

Age of the tree in years	FYM	N	P	K
1 Year	10	100	100	50
II	20	160	165	140
III	20	320	310	250
IV	25	500	380	500
V	25	600	400	520
VI Years and above	25	700	500	600

For shervaroyan hills (for trees above 6 years old)

700: 375:600 g/tree NPK along with VAM (*Glomus fasiculatus*) @ 1 kg/ tree. The manures and fertilizers are mixed and applied in broad ring formed at a distance of 80-100 cm away from the trunk at a depth of 10 cm, then closed with top soil and irrigated. In hilly areas where the pH is very low, depending upon the pH, 2 to 4 kg of lime or dolomite (not mixed with chemical fertilizer) should be applied for each tree once in two years, one month ahead of the application of regular fertilizers.

Micronutrient Spray

The following micro nutrients mixture can be applied during new flush: $ZnSO_4$ - 600 g, $MnSO_4$ - 600 g, $MgSO_4$ - 600 g and $FeSO_4$ - 600 g. By spraying the above micronutrient mixture in 450 lit of water the deficiency of micronutrients can be avoided. Application of Vesticular Arbuscular Mycorrhizae (VAM) which are symbiotic fungi @ 1 kg/tree will help to accumulate phosphorus zinc, copper and sulphur and by this quantity of 125 g of phosphorus can be reduces from the regular does of 500 g.

Fertigation

The fertigation treatment consisting of 500:240:70 g of NPK dose with 20% depletion of available water content (AWC) found best to increase higher canopy volume, fruit weight, TSS content, juice content and fruit yield in Nagpur mandarin (*Citrus reticulata*). Gypsum and neem coated urea (N 450 g/ plant/ year) found best to increase canopy volume and fruit yield in Nagpur mandarin. Application of 7.5 kg neem cake along with 600:300:600 g NPK per plant per annum produced maximum fruit yield and best quality fruits in Khasi mandarin. Two sprays with GA3 15 ppm + Urea 1% Benomyl 1000 ppm in April and May resulted in 22.20% reduction of fruit drop over control in Nagpur mandarin.

Irrigation

The frequency and quantum of irrigation depends upon many factors such as species, stage of the crop, soil type etc. Mandarins are mostly grown under rainfed conditions; however, irrigation during dry months is beneficial. During the periods of rapid growth, flowering and fruit set, the citrus trees are sensitive to moisture stress. Moisture stress during fruit development phases affects the size of the fruit besides causing excessive fruit drop. They are equally sensitive to excessive moisture and water logging conditions. Care should be taken to avoid the irrigation water directly contacting the tree trunk.

Crop Regulation

In central India, mandarins bloom thrice a year. The February flowering is known as Ambe bahar; June flowering as Mring bahar and October flowering as Hast bahar. Under suchcircumstances, plants give irregular and small crops at indefinite intervals. To overcome this problem and to get fruitful yield in any of the 3 flowering seasons', treating mandarin trees has been practiced which is called resting or root exposure or bahar treatment. In this method, roots of the plant are exposed too sun by removing up to 7 -10 cm soil around 40-60 cm radius of tree trunk. The water is withheld for a month or two month before flowering. As a result of water stress, leaves show wilting and fall on the ground. At this stage the roots are again covered with a mixture of soil and FYM and irrigated immediately.

Subsequent irrigations are given at suitable intervals. Consequently, plants give new vegetative growth, profuse flowering and fruiting. However, in light sandy and shallow soils, exposure of roots should not be practiced and mere withholding of water for 2-3 weeks is sufficient for wilting and debilitation of trees. It depends upon the choice of the grower as to which of the 3 bahars is to be taken to get maximum profit. As the availability of water is a problem in central India during April – May, the farmers prefer mrig bahar (June) so that the plants are forced to rest in April – May. Resting treatment is not feasible in North India, as mandarin plants normally rest in winter and flower once a year. It is experienced that resting treatment in general is a devitalizing process and should be resorted to only under the advice and direction of a technical expert.

Growth Regulators

To increase the fruit retention spraying the trees at flowering and again at marble stage with 2,4-D at 20 ppm or NAA 30 ppm.

Training and Pruning

The water shoots and rootstock sprouts should be periodically removed. Trees are trained to single stem with 4 – 6 well – spaced branches for making the basic framework. Further no branches should be allowed from the trunk up to height of 45-50 cm from the ground level. An ideal mandarin tree should be low headed with dome like crown. The bearing trees require little or no pruning. Pruning of bearing trees consists of removal of dead, diseased, criss-cross and weak branches. Removal of water shoots and suckers of rootstocks is also highly essential. Pruning of non-bearing trees can be done at any time of the year, but for bearing trees, the best time is after harvesting, during late winter or early spring when these are in somewhat dormant stage.

Intercultural Pperations

Heavy bearing branches should be supported by bamboo stakes. During summer, the tree basins should be mulched with dried leaves so as to prevent moisture loss. Dry grass mulch is very effective in maintaining soil moisture, soil temperature and reducing fruit drop.

Harvest and Yield

A small crop can be obtained from 4 year old tree and the yield will be higher from 7th year. From flowering to fruit maturity it takes 9 months. In Coorg, the main crop is harvested in December to April. A small off-season crop is obtained during July to September. In Tamil Nadu, the main season of harvest is November – December. Harvest should be done at right maturity. About 1000-1500 fruits can be harvested from a tree per year. The fruits harvested at green mature, colour break and ripe stage dipped in 8 per cent wax retained the freshness of fruits upto 60 days under ambient condition. Kinnow mandarin fruits wrapped in HDPE can be safely stored upto 60 days and 80 days at ambient and cold storage respectively without much loss of quality.

CITRUS FRUITS - PHYSIOLOGICAL DISORDERS

Fruit Drop

Citrus drops flowers and fruits at different stages as a natural means of adjusting to it resources. These are normal and may not affect the yield. Pre-harvest drop of well grown fruits prior to maturity is a serious problem reducing the yield. The main causes may be due to Climatic factors, Improper water management. Lack of nutrition and Relation of seed to fruit drop.

Control: Spraying 10 ppm 2, 4-D, 2 months before harvesting. Fruit drop may be due to presence of pests and diseases also which can be controlled by spraying 1 % Bordeaux Mixture.

Fruit Cracking

It is due to sudden changes in temperature and also due to moisture stress condition. Cracking of fruits may be radial or transverse. Secondary infection is also possible due to Aspergillus, Fusarium or Alternaria

Management

Apply light irrigation at frequent intervals and Application of potassium during fruit development.

Granulation

The juice vesicles become hard, enlarged and turn opaque grayish in colour. The density of pulp is increased, juice contains increased minerals (Calcium, sodium, potassium) and decreased carbohydrate and organic acid. It results in lignification of juice cells that leads to formation of sclerenchyma. High humidity and fluctuation in temperature are the major factors.Young trees are more prone to granulation than older trees.Application of more nitrogen, excess irrigation, large size of fruits, rootstocks are also a cause. Mandarins on jattikhatti rootstock are more susceptible than sweet orange .

Management

Avoid excess moisture, Spray lime @ 20kg in 450 l of water. Spray zinc (0.5%) and copper (0.5%).

Sunburn or Sunscald

The portion that is exposed to sun develops yellow patches which turn brown and become hard. The inner portion becomes dessicated and discoloured.Affected fruits are malformed and have low juice content. Severely affected fruits drop off and leaves turn brown.

Management

Spraying lime solution @ 20g//l before summer. Regulation of irrigation to reduce the temperature. Mulching the tree basins. Incompatibility of rootstock and scion, salinity, water logging and mismanagement of citrus orchard are causes to citrus decline.

CITRUS QUALITY IMPROVEMENT

Degreening in Mandarins

Although mandarins may attain optimum maturity standard but the fruits may not be attractive at the time of harvesting due to lack of good yellow colour. Accordingly, degreening of mandarins with the application of ethrel (50 ppm) one week before the harvesting develop golden yellow colour within 5 days of the treatment.

Colour Development in Sweet Orange

Colour development in sweet orange especially under tropical condition is a problem and to obtain colour on fruit, 1000 ppm of ethephon as a pre-harvest spray or fruit drip in combination with Benlate 1000 ppm is recommended.

❑❑❑

CHAPTER

10

GRAPES

Botanical name: *Vitis vinifera*

Chromosome no.: 2n = 38

Family: Vitaceae

Origin: Armenia

The grape is one of the important fruit crops grown in India. It is a native of Armenia (USSR) and was introduced in India during 11th century and in Tamil Nadu it was introduced in Madurai and Salem districts in 1832 by French Missionaries. The area uder grapes in India have been increased substantially in the last four decades and the present area is about 0.25 lakh ha with an annual production of 4.08 lakh tones. The principal centres of grapes cultivation are Madurai in Tamil Nadu, Nasik and Poona in Maharashtra, Hyderabad and Aurangabad in Andhra Pradesh. Punjab and Haryana. In Tamil Nadu until recently grapes were grown only in two districts viz. Madurai and Dharmapuri. Now it has been extended to other districts like Coimbatore, Salem, Tiruchirapalli and Tirunalvelli. Cultivation of grapes is also called as **Viticulture**

Nutritive Value

The fruits are a rich source of carbohydrates (13.1%), protein (0.6%) calcium (20mg/100g), phosphorous (25mg/100g) and iron (0.5mg/100g).

Climate and Soil

The grapevine is a subtropical plant and in temperate regions the vines are deciduous while in the tropics the vine is evergreen, growing continuously. But It originated as a temperate plant only by subsequently acclimatized to subtropical and tropical conditions also. For best development, the vinifera grapes reqires long warm to dry summers and cool winters in temperate regions. The grapevine in

Tamil nadu grows luxuriously without an unproductive dormant bud, characteristic of most vineyards in the temperate and subtropical regions. The grapes growing districts receive moderate rainfall distributed in such a manner that it done no harm to any phase of the crop. The rainy seasons are broken by frequent spell of dry weather so that the vine receives heat and sunshine sufficient to mature two crops satisfactorily. The significant climate feature of Tamil Nadu which in contrast to northern grape belts enables to harvest even five crops in two years against only one crop in other parts of India. Further, hail storms, gales an d floods are practically unknown in the grape districts of Tamil Nadu.

The climate in South India such as Bangalore in Karnataka, Dharmapuri and Theni districts of Tamil Nadu is slightly humid and tropical. Here the maximum temperature goes upto 35°C and the minimum temperature does not fall below 12°C due to warm winter condition there is practically no rest period. Almost rainless period during November - June favours heavy as well as sweet crop. Here the vines are pruned twice. Early December pruning yields a sweet crop during April and summer pruning (May) yields a slightly sour crop during September.

The weather conditions at the time of pruning and following pruning and upto berry development should be free o heavy rain as continuous cloudy and rainy weather, besides interfering with flower and fruit set, encourages pests and diseases. In tamil Nadu, pruning is done in such a manner that vines are not in flowering during October and November, the Northeast monsoon seasons. Grapes can be cultivated in a wide range of soils, from low to high fertility. An ideal soil for grapevine should be well drained, rich loamy one with low water table. The soil depth should be at least 60 cm

Varieties

The existing cultivars can be broadly grouped under the following categories.

Commercial varieties

Coloured seeded - Bangalore Blue and Gulabi (Muscat)

Coloured seedless - Beauty Seedless and Sharad Seedless

White seeded - Anab-e-Shahi, Dilkhush (clone of Anab-e-Shahi)

White seedless - Perlette, Pusa Seedless, Thompson Seedless, and its clones Tas-AGanesh, Sonaka and Manik Chaman

a. ***Table grapes:*** Utilized either as a fresh fruit or for decorative purpose. They must have an attractive appearance, good eating quality, good shipping and storage qualities e.g. Thompson Seedless, Beauty Seedless, Bangalore Blue.

b. ***Raisin grapes:*** grapes which produce an acceptable dried product are called raisin grapes. Seedless cultivars with high sugar content production yellowish or greenish raisins are preferred. E.g. Black Corinth, Sultan, Thompson seedless.

c. ***Juice grapes:*** The juice of some cultivars with muscal flavour produces an acceptable unfermented beverage when it is preserved by pasteurization other means, E.g. Bangalore blue, Beauty Seedless.

d. ***Wine grapes:*** Cultivars which have high sugar content and low acid with can produce satisfactory wine is called wine-grapes. E.g: Muscat Canelli, Black champa.

e. ***Canning grapes:*** seedless cultivars with white grapes with larger barriers are preferred for canning. E.g. Thompson seedless, Pusa seedless, Perlette

A. SEEDED VARIETIES

Anab-e-Shahi

This variety had originated from a bud sport. The yield is 75-90 tonnes/ha per year. The bunches are large with large attractive oval berries which are pale green in colour. The keeping quality is excellent. The TSS is 12-16°brix and acidity is 0.4 to 0.5%. It is a late maturing variety.

Bangalore Blue (Bangalore Purple)

Important variety around Bangalore used in brewery industry and also for making juice. Medium yield, bunches hardly shouldered, compact, purple in colour with uniform ripening. TSS 16-18°brix, pulp pale green. Acidity 0.8 to 0.9%, good keeping quality. Highly resistant to anthracnose and powdery mildew diseases. It possesses labrusca blood i.e. it is a natural hybrid between V. vinifera x V. labrusca. This variety is suited for kniffin and bower systems of training.

Cheema Sahebi (Selection No.7)

This is a seedling selection from open pollinated progenies of Pandhari Sahebi at Agricultural College, Pune by Dr.G.S.Cheema. The vine is vigorous with heavy yield.

Paccha Draksha (Bhokri)

It is grown in Maharashtra and Tamil Nadu. Vigorous vines, heavy yield, bunch medium to large, compact, berries green, medium size, round to slight oval in shape, TSS 17-18°brix.

Black Champa

It is a selection developed at the Indian Institute of Horticultural Research (IIHR), Bangalore. The vines are vigorous and the yield is medium. The high quality purple coloured berries are highly suitable for table, juice and wine making.

Champion

The juice of this variety is highly flavoured having a TSS 21°brix and suitable for juice industry. It has small bunches with medium sized berries.

Early Muscat

This variety is medium vigorous and perform well on kniffin system of training. The fruits are green with good Muscat flavour.

Gulabi (Panneer Drakshai)

It is otherwise called Karachi or Black Prince. This is the most important cultivar for Coimbatore conditions. Vine is very weak, having cane with basal fruitful buds, pruned to short canes (5-6 buds), high tolerance to powdery mildew and anthracnose.

Dilkush

It is a bud sport of Anab-e-shahi producing golden yellow enlongated berries in attractive bunches.

Grape hybrids and their specific characters

Hybrid	Parents	Distinguishing characters
Arkavati	Black champa X Thompson Seedless	Good for raisin making
Arka Kanchan	Anab-E-Shahi X Queen of Vineyard	Late maturing and good quality
Arka Hans	Bangalore Blue X Anab-e-Shahi	Suitable for white wines, berries with poor attachment
Arka Shyam	Bangalore Blue X BlackChampa	Good for juice and wine making, suitable for double cropping in S.India
Arka Neelmani	Black champa X Thompson Seedless	Table as well as red wine making
Arka Shweta	Anab-e-Shahi X Thompson Seedless	Seedless and suitable for table purpose

[Table Contd.

Contd. Table]

Hybrid	Parents	Distinguishing characters
Arka majestic	Anab-e-Shahi X Black Champa	Table purpose
Arka Chitra	Angoor Kalan X Anab-e-Shahi	Table purpose
Arka Soma	Anab-e-Shahi X Queen of Vine Yard	Suitable making good quality wine making
Arka Trishna	Bangalore Blue X Queeen of Vineyard	Suitable for wine making
Arka Krishna	Bangalore Blue X Convent large Black	Suitable for juice
Arka Urvashi	HurXBeauty seedless	Suitable for juice and wine.tolerant to Anthracnose
Pusa Navarang	MadeleineAngavine X Ruby Red	Tenturier,suitable for juice and wine

B. SEEDLESS VARIETIES

Thompson Seedless

It is also called as 'Sultanina' or 'Oval Kishmish' and is an introduction from the University of California, USA. It is commercially cultivated in Tamil Nadu and Maharashtra. TSS is 22-23°brix. This variety is suitable for bower and trellis systems of training. The berries are used for raisin making.

Pusa Seedless

This variety is a selection from Thompson seedless developed at IARI, New Delhi. Pusa seedless is similar to Thompson seedless in many characters except the shape of the berry which is round in this variety. It is used for table purpose and raisin making.

Perlette

It is a hybrid between Scolokertekhiralynoje 26 x Sultanina Marble The berries are medium in size, spherical and whitish green. Perlette has good keeping quality.

Delight

It is a sister seedling of 'Perlette'. An early ripening variety with good Muscat flavour. It is a good table variety with good shipping quality.

Beauty Seedless

This variety is medium vigorous in growth. It is susceptible to pre-harvest berry drop. Beauty seedless is suitable for head system of training.

Sarath Seedless

It is a selection from Beauty Seedless made by grapevine growers.

Kish Mish

'Kishmish Beli' is a variety with berries which are small sized, elongated and golden yellow. 'Kishmish Charni' is with brick red.

TABLE GRAPES

Arka Shweta (E-11/29)

A cross between Anab-e-shahi and Thompson seedless. Each bunch weighs 260 g, greenish yellow coloured obovoid uniform seedless berries. TSS 18-19°brix , acidity 0.5-0.6%. On bower system at 3 m x 3 m planting distance, it yields 31 tonnes/ha; comes to harvest in 153-155 days after pruning.

Arka Majestic (E-14/20)

A cross between Angur Kalan and Black Champa. Suitable for head system of training. Hence, 2 crops can be taken in a year. All buds are fruitful. It yields 38 tonnes/ha and 34 kg/vine on bower system, requires 160-163 days for harvest from pruning. Deep tan coloured uniform bold round berries, TSS 18-20°brix, each berry has 2-3 small seeds.

Arka Chitra (F-5/4)

A cross between Angur Kalan and Anab-e-Shahi. All buds are fruitful requiring no specific pruning, bunch weight 310 g. bunches are golden yellow with pink blush, slightly elongated berries each weighing 3.18 g, TSS 20-21°brix. On bower system spaced 3 m x 3 m apart, it yields 38 tonnes/ha and 34 kg/vine. From pruning to harvest, it takes 154-158 days.

WINE GRAPES

Arka Soma (E-9/3)

White berries, seedes, meaty pulp and Muscat flavoured.

Arka Thrishna (E-21/28)

Deep tan coloured berries, seeded with very sweet pulp, male sterile.

Propagation

Grapes are commonly propagated by rooted stem cutting or hard wood cutting; 20-30cm long cutting are made from ripened wood of the previous season. The lower cut is made just below a node. The cuttings prepared should have atleast 3 to 4 buds. The cuttings are planted in the nursery with one or two internodes exposed. Cuttings are soaked in 500 ppm of IBA for 12 hours (or) 12,000 ppm for 3 seconds. It gives good percentage of rooting. Grafting and budding is practiced with a particular root stock for specific requirement. Well matured healthy cutting form the roots within 3 to 4 weeks and strike roots to the extent of 90%. After a period of three months, the rooted cuttings could be potted in the tube pots. Grapes are also propagated by budding (chip budding) and grafting (whip, cleft, side grafting) so as to establish vines on a rootstock resistant to soil borne pests, disease, nematodes or alkaline or saline conditions.

Rootstocks

Rootstocks are employed for grapes to overcome salinity, nematode damage and to impart vigour to vines.

a. Phylloxera resistant root stock - Vitis *riparia*, *V. rupestris,* Teleki 5-A and St.
b. George.
c. Nematode resistant root stock - 1613, Dogridge, Salt creek, Freedom and
d. Harmony.
e. Saline tolerant - 1616, 1613Solanis, Ramsay, Dogridge and
f. Rupestris du hot.
g. Drought tolerant - 110 Richter, 140 Ruggeri, 99 Richter and 1103
h. Panlsen.

Preparation of Main Field and Planting

Trenches of 0.6 m width and 0.6 m depth are dug at a distance of 3 m apart for Muscat. Other varieties $1m^3$ size pits are dug at required distance recommended for different varieties Well decomposed FYM or compost or green leaf manure has to be applied in the trenches or pit and then covered with soil. Rooted cuttings are to be planted in the centre of the pit during June - July. In Tamil Nadu, they are usually planted from June to January.

Planting

In Tamil Nadu, they are usually planted from June to January. The rooted cutting is usually planted in pits. Pits of 1m x 1m x 1m size are dug at required distance recommended for different varieties

Variety	Spacing
Muscat	3 x 5 m
Pachadraksha	5 x 5 m
Anab-e-Shahi	6 x 6 m
Thompson Seedless	5 x 5 m

Well decomposed farm yard manure or compost, red earth and sand should be mixed in equal proportion and are used to fill up the pits. Rooted cuttings are to be planted in the centre of the pit.

Nutrient Management

A grapevine removes an appreciable quantity of nutrients from soil both in terms of wood and fruits. It has been estimated that an average crop of grape removes from the soil 40- 60 Kg N, 10-15 Kg and 50-70 Kg K per hectare. Hence, to maintain the soil fertility for consistent yield, it becomes necessary to replenish these nutrients through proper manures and fertilizers. The following manures and fertilizers are recommended for grapes in Tamil Nadu.

S.no	Nutrients	First year	Yearly increase	Full grown vine
1	Farm Yard manure	50 kg	25kg	200kg
2	Green leaf manure	50kg	20kg	200kg
3	Ammonium sulphate	500g	250g	1kg
4	Super phosphate	500g	250g	1kg
5	Muriate of potash	500g	250g	1kg
6	Bone meal	-	500g	2kg

The manures and fertilizers are applied to the soil soon following pruning removing the soil to a depth of 15 cm leaving a space of 30 to 60 cm around the trunk of the vine. Muriate of potash is applied in two split doses as it will improve the qualities of the berries. Half the dose is applied after pruning and the other half 60 days later.

Irrigation

During growing season, there should not be any water stress. Normally irrigation will be required once in 10 days in non-monsoon, but during monsoon rains, irrigation need not be given. During new growth, flowering and early and mid fruiting phase, there should not be any moisture stress. About two weeks prior to harvest, the irrigation should be stopped which will facilitate ripening of fruits. The vines should be irrigated soon after pruning and fertilizer application. Until the berries attain pea size, water is given at intervals of 5 to 7 days and thereafter until the commencement of maturity at 10 days interval. Irrigation is usually withheld until harvest which will help to improve the quality of berries.

Weed Management

Weed management is essential especially in young vine yards. Pre emergent herbicides should be applied after pruning of vines and before the monsoon. Several herbicides like simazine, diuran, nitralin, triflurali, paraquat, dalapon etc., are in use fore control of weeds in vine yards.

Training and Pruning

Proper methods of pruning and training contribute towards higher production of better quality fruits in grape. Training mainly concerns with giving the form and the direction of the trunk and arms and the position of the shoots. Training determines the form while pruning effects the functioning of the vine. It is done to concentrate the activity of the vine to the parts left after pruning. Pruning is the most important operation for the maintenance of fruitfulness and quality along with vigour of the vine. Before actually discussing the subject of training and pruning it is necessary to understand the various terms commonly used in these operations to make the subject more intelligible.

Trunk	The main stem of the vine which is vertical
Arms/Cordons (Primary)	The main branches arising from the trunk or extensions of the trunk usually grow vertically
Arms/Cordons (Secondary)	The branches arising from primary arms or extensions of the primary arms
Head	The region of the trunk from which the arms or canes arise
Shoot	The young growth(herbaceous) of the current season developing from a bud situated on the arm or trunk
Cane	The matured shoot of the past season
Spur	The shortened cane or part of the cane left after pruning
Fruiting spur	The spurs having a few buds some of which (usually the apical ones) sprout and grow into fruiting shoots.

Training

In the natural habitat, a grape vine is robust climber but it can be trained on any fashion. Although a no. of training systems are known only four namely bower, kniffin, telephone trellis, and head system are followed in India.

Head System

This is the cheapest and easiest system of training grape vines. In this system the vines are trained like dwarf bush. Less vigorous varieties and varieties producing fruitful shoots from the basal buds are suitable for this system. Ex. Beauty seedless, Perlette, Delight and Gold. In this system the plants are spread very closely to accomodate2000-2500 plants per acre. The vine is allowed to grow to a single stem with the help of stakes. After attaining a height of 3‘ the plant is topped and two lateral branches are encouraged. The plant is again topped at 4‘ height by which two or more laterals are developed. After keeping 4 laterals, 4‘ above the ground in all directions, the rest of the shoots are thinned out. These later cut to two buds at the first dormant pruning, will produce secondary arms. Generally two arms of about 20-30cm are kept on each lateral. At the time of second pruning, normally 1-2 fruiting spurs are kept on each secondary arm. After 3-4 years, the vine becomes like a dwarf bush and needs no stake.

Advantages

- Simplicity in shape
- Ease in training
- In expensive to establish
- Possibility of cross cultivation

Dis-advantages

- The vines are slow to come to full production
- Increased possibility of bunch rot and poor colour
- The bud and flower drop is maximum compared to other systems.

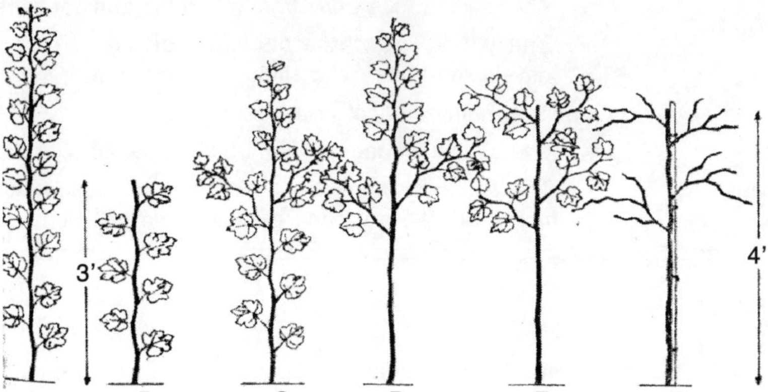

Fig. 16. Steps in training a grape vine on Head System

Pandal System

This system is also called as Arbour, Pergola, Mandwa, Over head or Bower system. Owing to the vigorous growth of the vine and pronounced apical dominance in the tropics, this system is most suitable for many of the commercial grape cultivars. This is more popular system for Anab-e-shahi in A.P.This is best suited for vigorous varieties, which don't perform well on other systems.Inspite of being the most expensive; this is being adopted on a large scale almost in all the grape growing regions of India.

In this system the vines are spread over a criss cross network of wires usually 7' (2.1m) above the ground supported by pillars(Concrete, stone or iron).Galvanized wires of 5,8 and 10 gauge thickness and turning buckles are used. Only the best growing shoot from the plant is allowed to grow upright along the stake provided up to the bower height. When the vine reaches the wires, it is pinched off 15cm below the pendal level to facilitate production of side shoots close to the wires. Two vigorous shoots in opposite direction are selected and allowed to grow in opposite directions on the wires overhead. These two shoots develop into primary arms. On each primary arm three laterals on either side at a distance of 60cm (2') (along the wires) are kept as secondary arms. Thus, there will be 12 secondary arms on each, which after maturity form fruiting canes. These primary and secondary arms for the permanentframe work of the vine.

The vines are allowed to trail straight along the wires by tying intermittently with banana fibre.

Advantages

- Greater spread of the vines.
- Better exposure of the foliage to the sun, resulting in better maturity of the canes.
- Higher production.
- More uniform bunch colour
- Superior quality of fruits, which are free from sunscald.
- The vines in general give uniform performance
- It gives good protection to the canes against hot desiccating winds with ease in bird scaring.

Dis advantages

- It is the most expensive than all other systems
- Pruning, training and spraying operations become difficult
- The spraying material cannot reach effectively the leaves and shoots.

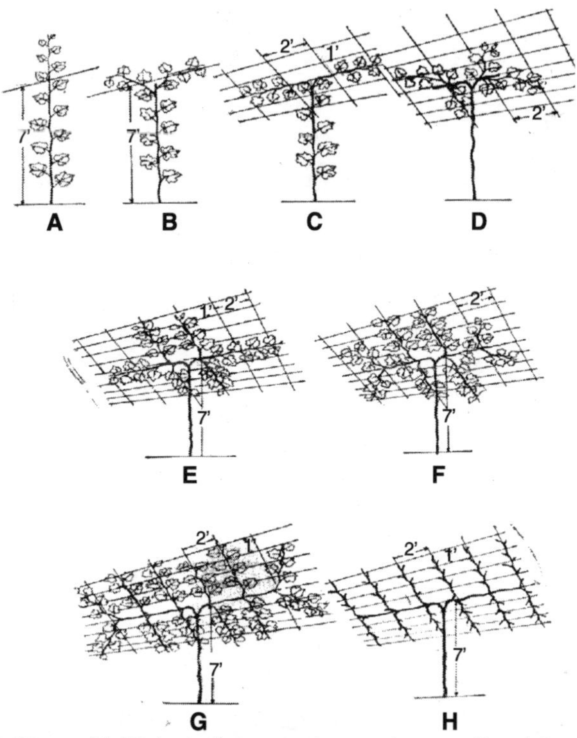

Fig. 17. Steps (A-H) in training a grape vine on Pendal system

Telephone System (Overhead trellis / Telephone trellis system)

This system is suitable for moderately vigorous varieties with more apical dominance. The chief demerit of kniffin system, where the lower arms are rendered unproductive, is modified in this system by wires, stretched at one height like telephone wires. It is also an improvement over bower system in respect of ventilation and light penetration. It is relatively less expensive than kniffin system. The usual spacing provided for each vine is 3x3m. Trellies are erected by using the granite stone pillars of 8' length and 6 x6 thick at the ends and 8 x4 x4 in the middle of the lines. The middle pillars may be spaced at 20' distance. Cross arms of 41/2' length are fixed on each pillar at a height of 5'.These can be iron blades of 4 width and ¼ thick or the angle iron pieces of 2 x2 width and¼ thick. Three wires of 8 gauge thick galvanized iron are pulled horizontally over the cross arms at a regular spacing of 2' using turning buckles at the end of pillars are supported side ward. In this system the vines are allowed to grows straight up to a height of 1.5m (5') and then trained over head on a canopy of usually 3or 4 wires (45-60cm apart) fixed to the cross angle arms supported by vertical pillars or posts. The young growing vines are supported by bamboo sticks. After reaching the height of telephone (5') the tip should be pinched off to encourage side shoots close to wires. Two vigorous side shoots (cross to wires) are selected as primary arms from which four vigorous laterals on each side along the wires are allowed to develop on secondary arms. Each complete secondary arm can carry 6-8 fruiting units.

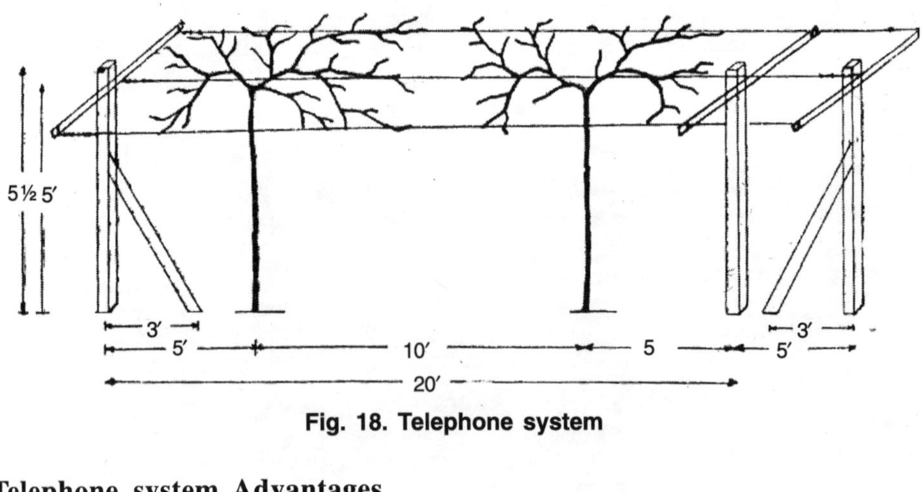

Fig. 18. Telephone system

Telephone system Advantages

- Greater spread of the vine
- Better exposure of the foliage to the sun resulting in the better maturity of canes.

- Higher production
- More uniform bunch colour
- Superior quality of fruits free from sunscald
- Vines in general give more uniform performance.

Dis advantages

- Cost of establishment is high. At present rates it may cost 60-70 thousand rupees.
- It is a system difficult to develop i.e. vine training needs a lot of skill and effort.
- The bunches are not as well exposed to light as kniffin system.
- Not suitable for vines making low to moderate growth.

Kniffin System

It is also called as espalier system. It is a system of training grape vine in which the arms of the vine are tied to horizontal wire at the same level above the ground. This system is not as common as the bower system. It is suitable for the moderately vigorous varieties with less apical dominance. Closer planting is adopted for this training system with in the row and 3m (10') between the rows. Galvanized iron wire of 8 gauge thickness is stretched parallel to the ground at a height of 75cm above which two or more wires are stretched at successive heights of 60cm.when the plant crosses the first it is topped leaving a bud above the wire. Two laterals are developed on either side of the plant along the wire and the terminal shoot is allowed to grow vertically. Similarly, a pair of laterals is developed along the second and third wire. Thus each vine will have six arms. In some cases only two pairs of laterals are developed at a height of 41/2' (1.35m) and 61/2' (1.95m) from the ground level and in such cases it is called four arm kniffin system. This system is suitable for Beauty seedless, Early Muscat, Bhokri and Delight.

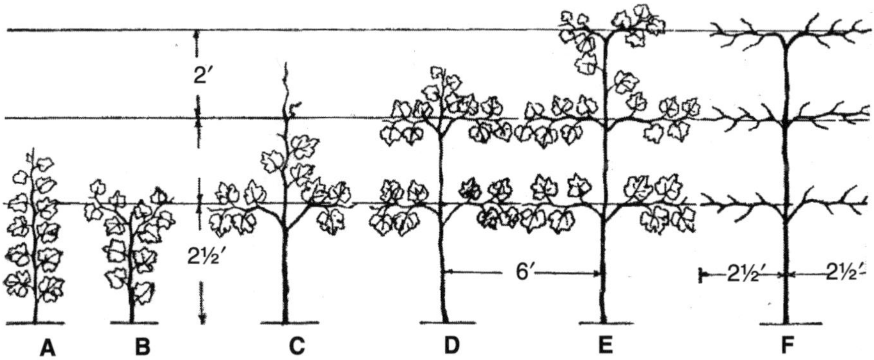

Fig. 19. Steps in training a grape vine on Kniffin system

Advantages

- This system is cheaper when compared with bower system.
- It is a good system for obtaining full crop on vines, the basal buds on the canes of which are sterile and which require long cane pruning.
- It is good for small clustered varieties which require fruit thinning for improved quality.
- The system allows more lateral spread of fruit bunches than cordon system.
- The average yields are 10-15kg per vine when planted at 3.0 mx3.0 m spacing.

Dis-advantages

- Cultivation is possible in one way
- The lower arms become unproductive after some years.
- The arms produce fruiting wood mainly at the extreme ends only.
- From the pruning point of view, it is very difficult and exacting system. Since the retained fruiting units are very few, they must be perfect and pruner must have a thorough knowledge for their appropriate selection.
- Its cost of establishment is about 50-60 thousand rupees per hectare.

This system is mainly confined to research institutes and it is not popular with the farmers in any of the regions of the country.

Pruning

In grape pruning is done only once in North India during the month of January to make the fruitful buds to sprout but in south India, pruning is done twice in a year, once in summer and again in winter. Grape vines in these regions grow continuously without any dormancy (due to tropical climate).Hence by pruning in April (summer) the vines are forced to have a rest period, which helps in fruit bud differentiation. Pruning time mainly depends on rainfall and temperature. Pruning is adjusted so that there is no coincidence of rainfall with fresh growth and flowering and also winter doesn't set in with in 8-10 days after pruning.

Pruning Refers to the Judicious Removal of any Plant Part

- To establish and maintain desired vine shape
- To increase productivity
- To facilitate various cultural operations
- To distribute proper amount of bearing wood over the vein for consistent productivity

Summer Pruning

It is done during March-April in the states of A.P. and Karnataka, but in July in Tamil Nadu.In this pruning the canes are cut back to one or two bud level for building up the fresh vegetative growth. Hence it is called **back pruning or growth pruning**.

Winter pruning: This is done during the last week of November in A.P. and Maharastra, during the second and third weeks of October around Bangalore, but at any time of the month of October in the interior northern districts of Karnataka and in December in Tamilnadu. The mature canes (about 6 months old) are pruned. Entire foliage and immature shoots are removed. Levels of pruning differs with varieties.Anab-e-shahi and Bhokri are pruned to 5 bud level, Thompson seedless to 10 buds, Bangalore Blue to 4 buds and Gulabi to 9 buds. This pruning is also called as **forward pruning**.

Some of the varieties like Perlette, Beauty seedless, Bangalore blue, Bhokri etc. produce fruits on the shoots arising from the basal buds on the cane. In such varieties the canes are headed back to 4-5 buds. Such varieties are called Spur pruned varieties. On the other hand the Pusa seedless, Thompson seedless varieties in which the fruitsare produced on the shoots arising from terminal buds, the canes are headed back to 8-12 buds. Such varieties are called Cane pruned varieties.

Cropping

Grape has two kinds of functional buds viz (1) leaf buds and (2) fruit buds. The leaf buds is a rudimentary sterile shoot and when it opens it will not bear fruit. Fruit bud contains a shoot having both rudimentary leaves and flower clusters. The fruit buds take 47 to 70 days from bud break to blooming.

Special techniques to improve fruit quality

a. After fruit set, tipping of the shoots and the clusters are to be done.

b. The tendrils in the shoots are to be removed.

c. Nipping the growing shoots of axillary buds and terminal buds at 12-15 buds.

D. THINNING THE BUNCHES

Excessive bearing of vines can be reduced by cluster or berry thinning. This helps to increase the size of berries and also the quality. Thinning should be done when the berries are of pea-size in varieties producing compact bunches. The degree of thinning is limited to 20 per cent of the berries in the cluster.

E. Girdling

Ring of bark removed from any part.

- 0.5 cm wide ring. Done at several stages for several purposes.
- One week before bloom - To improve berry set.
- Just after berry set - To increase berry size.
- At the beginning of color change
- To advance ripening
- To obtain uniform colour
- Better eating quality

F. USE OF GROWTH REGULATORS

Application of gibberellic acid (GA_3) at 50 ppm, - at the just opened flowers in the clusters (calyptras stage) increases their size. Besides, ethrel is used to advance ripening by a week and improve quality including colour improvement and uniformripening. It is normally applied at 250 to 500 ppm concentration at berry start ripening stage.

Use of Plant Growth Regulators

Encouraging responses were observed both in seeded and seedless varieties of grape by the use of growth regulators. Effect on fruit set: A good fruit set was obtained by spraying the flower cluster thoroughly 4-6 days after full bloom with 100ppm Gibberellic acid or 20ppm Parachloro phenoxy acetic acid. This increased set in current grapes, Thompson seedless and black Corinth. Effect on berry size: Increase in berry size in Anab-e shahi, Kishmis and Bhokri varieties was reported when GA was applied at 40ppm at bud and flower stages. Higher concentrations resulted in the increase in the length of berries.

Effect on Cluster Size

Use of GA, TIBA and PCPA resulted in lengthening the cluster parts especially the pedicles.

Effect on Maturity

With the application of Benzothiozal A-Oxyacetic acid (BOA) maturity can be regulated. Maturity can be delayed by 15 days with the application of this Benzothiozal A-Oxyacetic acid.

Harvesting

Bunches of most of the cultivars are ready for harvest between 100 to 120 days after pruning. It has to be harvested at proper stage for quality. Generally the berries are to be harvested at ripening stage, characterized by showing colour and softening of berries. The sugar content increases at this stage. In seeded grapes, the seeds become dark brown when they are fully ripe, while in seedless varieties, their characteristic berry colour develops fully. Grapes are harvested when fully ripe, since they do not ripen after harvesting, so must be left on vine until they are fully ripe. While harvesting, the clusters should be handled only by the stem so that the natural appearance is not impaired due to ribbing of the cluster. The cluster should be removed from the vine by cutting with a sharp knife near its attachment to the cane. The fruits of grapes require specific amount of heat units over and above basal temperature for proper maturity and ripening. The heat requirement of most varieties of grapes ranges from 2900 to 3600 units.

Yield

Although grapevines yield in the 2nd and 3rd year of planting, good crops can be expected only after the third year. Yields of grapes are highly varied as much as 12.5 to 21 tonnes as early as 10th month from planting followed by equally heavy crops every six months thereafter until about 10 years. Pachadraksha yields 35 to 37 tonnes in about 18 months followed by 20 tonnes within another 6 months. This continues over a number of years. In the case of Anab-e-shahi, the cropping is unpredictable ranging from few bunches to 25 tonnes per hectare. The yield ranges from 20-45 tones/ha depending on the variety and system of training. The yield potential of grape in India is highest in the world. Grape variety Anab-e-Shahi has recorded yield as high as 92 tonnes/ha, whereas Thompson Seedless has 48 tonnes/ha. The average yield of Anab-e-Shahi and Bangalore Blue is 40-50 tonnes/ha, while that of seedless varieties is 20 tonnes/ha.

POST-HARVEST TECHNOLOGY

Packing

The bunches are placed in the pack in such a way that the stem end lies on the outer side of the package which facilities easy removal of bunches during retail marketing. The CFB boxes of 2 to 4 kg are very popular for packing grapes. "Grape Guard" is a brown-coloured paper (which is impregnated with potassium sulphite) which releases sulphur dioxide gas when it gains moisture. Sulphur dioxide helps checking the grey mould disease of grapes which occurs at ambient as well as under cold storage conditions. It also helps keep the stem of the bunch from drying up. For getting the best advantages from "Grape Guard", the grapes should be wrapped in thin tissue paper or perforated polythene bags and a strip of Grape Guard placed in it before packing the grapes in box.

Storage

Shipping of table grapes is done immediately after harvest when possible. The grapes meant for storage should be packed and pre-cooled soon after harvest. Therefore, the grapes to be cold stored should be pre-cooled to 0-2°C using forced air-cooling and shifted to the cold store immediately after harvest. The grape berries can be kept without spoilage for 7 days at room temperature. Grapes can economically be stored up to 40-45 days in cold storage. The optimum storage temperature recommended is 2 to - 1.5°C. Fresh grapes can be stored for 90 to 180 days of time at 0-1°C and 95% RH depending on cultivar. Raisins can be stored for long periods of time at room temperature due to low moisture content.

PHYSIOLOGICAL DISORDERS

Pink Berry Formation

It is a common disorder in Thompson Seedless and its clone Tas-A-Ganesh in Maharashtra. Pink blush develops on a few ripe berries close to harvesting. The pink colour turns to dull red colour and the berries become soft and watery. They do not stand for long after harvesting. Although the definite cause of the disorder is not known, it is recommended to spray a mixture of 0.2% ascorbic acid and 0.25% sodium diethyl dithiocarbamate at fortnightly intervals commencing berry softening.

Short Berry

The berries become underdeveloped and undesirable in the bunch. This is due to the formation of very large number of bunches on the vine indicating higher crop load. Dipping the bunches at berry stage in ethephon @ 25 ppm for 1 min. will control this disorder.

Uneven Ripening

Presence of green berries in a ripe bunch of coloured grapes is called uneven ripening. It is a varietal character and a problem in Bangalore Blue, Bangalore Purple, Beauty Seedless and Gulabi grapes. Within a variety this problem varies from bunch-to-bunch. Generally inadequate leaf area and non-availability of reserves to a developing bunch is the reason. Cultural practices like cluster thinning, girdling and use of growth regulators can reduce uneven ripening. Application of Ethephon (250 ppm) at colour break stage is recommended to reduce the problem.

Post-harvest Berry Drop

This is due to weak pedicel attachment to the berries. This is common in Anab-e-Shahi, Cheema Sahebi and Beauty Seedless. Spraying of NAA (50 ppm), a week prior to harvesting can minimize the post-harvest berry drop.

Flower-bud and Flower Drop

When panicles are fully expanded, the flower-buds drop before the fruit set. This is common in north India but not in the south. The reasons for this disorder are not known. Stem girdling about 10 days prior to full bloom can reduce the problem.

❑❑❑

CHAPTER 11

PAPAYA

Botanical name: *Carica papaya* L.,

Family: Caricaceae

Chromosome no.: 2n = 18

Origin: Tropical America

The papaya is native of Tropical America was introduced to India in the 18th century. It is now grown in almost all tropical and subtropical countries in the world. It has originated in tropical America and was introduced from Philippines through Malaysia to India in the latter part of 16th century by Portuguese. It is one of the few fruit crops that flowers and fruits throughout the year giving early (9-10 months after planting) and high yields (about 100 tonnes per hectare). It is also called as 'Melon Tree'. Ripe papaya is yellow in colour due to the presence of 'caricaxanthin' pigment. Papaya is a wholesome fruit and is rich in vitamin A (2000 IU/100g). Papaya yield a valuable proteolytic enzyme called papain. Papaya is usually dioecious but hermaphrodite type and gynodioecious types are also recognized. In dioecious type, both male and female plants are separate. The male flowers are found on long pendulous panicle. Female flowers are solitary and are much larger than male. In gynodioecious type, the female and bisexual flowers are borne one on the same plant. The fruit develops from female flowers are globular in shape while fruits that develop from bisexual flowers are elongated in shape.

Nutritive Value

Papaya can ensure a good supply of vitamin A and C, which are essential for good health. Moreover it is a cheap fruit available in all seasons, in all places to all the people. Papaya is a wholesome fruit and is rich in vitamin A (2000 IU/ 100g).

Climate

Papaya is tropical in its climatic requirement. It is susceptible to frost. It requires warm humid climate and can be cultivated up to an elevation of 1000-1200 m above MSL. It is very sensitive to frost but withstands extremes of temperature. Temperatures below 10°C will affect the growth and fruit set. It grows well in regions where summer temperature doesn't exceed 38°C but it can stand up to 48°C but optimum range is 22-36°C, a day temperature of 35° C and 25°C night temperature are most suitable. It is adapted to a wide range of rainfall conditions ranging from 35cm to 250cm annually; however, excessive moisture adversely affects the crop as well as fruit quality. It does not stand strong (80Km/Hour) or hot winds. Dry climate during flowering often causes sterility while the same conditions during fruit maturity add to the sweetness of the fruit. Papaya performs well in tropical climates where summer temperature ranges from 35°C to 38 °C. At higher elevations, the fruit quality is usually lower. It cannot tolerate very hot summer or frost, this limits cultivation in Northern India. It cannot tolerate, very hot summer or frost, a dry warm climate tends to increase the sweetness of the fruits. In strong wind prone areas, wind breaks have to be provided to save the trees from wind damage. Tamil Nadu is an ideal home for growing papaya because of the mild temperatures and freedom from mosaic and leaf curl virus diseases. These features help all the year round cultivation of papaya.

SOIL

Papaya grows well under well drained soil with a pH range of 6-7. It prefers fertile soil, deep clayey and calcarious and rocky soils are not suitable. It does well in varied soil types, the best performance is observed on loams of uniform texture upto 1.8 m in depth. The most important requirement is that the soil should have good drainage. Even two to three cm of water stagnation around the tree for a few hours is likely to damage them due to the collar-rot disease occurrence. Under water stagnated conditions and in soils with poor drainage foot rot disease may cause heavy mortality. Hence, heavy soils should be avoided as papayas cannot withstand water stagnation for more than 48 hours.

Varietal identification

Based on the sex expression, papaya varieties can be either classified as dioecious or gynodioecious. The dioecious varieties produce male and female plants in 1:1 ratio when propagated from seeds. The gynodioecious varieties produce female and bisexual (hermaphrodite form) in 1:2 ratio.

The varieties CO 1, CO 2, CO 4, CO 5 and CO 6 are dioecious while CO 3 and CO 7 are gynodioecious. Among them, CO 2 and CO 5 are highly recommended for papain extraction while CO 3 and CO 7 suit very well for table purpose.

VARIETIES RELEASED FROM TNAU, COIMBATORE

CO 1 (1972)

It is a dioecious variety selected from 'Ranchi'. The plants are dwarf in stature and the fruits are round with flattened base. They are medium sized and smooth skinned. The fruits weigh around 1.2 to 1.5 kg with a TSS of 11-13° brix and have good keeping quality. Slight nipple and ridges are present at the apex. Fruits have orange yellow flesh with no papain odour and hence suitable as a good table variety.

CO 2 (1979)

It is a dioecious type evolved through pureline selection from a local type. Each fruit weighs about 1.5 to 2.0 kg and has a good papain yield (4-6 g from each fruit). The fruit contains 75 per cent pulp with a pulp thickness of 3.8 cm. Fruits are oblate and large in size, flesh yellow coloured, soft with TSS ranging from 11.4 to 13.5° brix. It is a dual purpose variety suitable for fruits and papain extraction. The yield ranges from 80-90 fruits per tree. The dry papain yield is 250 to 300 kg per hectare.

CO 3 (1983)

It is a hybrid derivative of the cross between CO 2 (female parent) and Sun Rise Solo (male parent). It is a gynodioecious type suitable as a table variety. The fruits are medium sized, red fleshed with a TSS of 14.6° brix. The mean fruit weight is 0.5 to 0.8 kg. Fruits are pyriform in shape. The variety yields 90-120 fruits / plant / year.

CO 4 (1983)

It is a hybrid derivative of the cross between CO 1 (female parent) and Washington (male parent). Petioles are purple coloured. Fruits are medium sized weighing 1.2 to 1.5 kg. It is a dioecious type. The fruits are yellow fleshed with a TSS of 14.3° brix. The variety yields 80-90 fruits per annum.

CO 5 (1985)

It is a selection from 'Washington' type with the characteristic pink petiole. It is a dioecious type suitable exclusively for papain production giving 14.45 g of dry papain per fruit. The variety yields 75-80 fruits per tree and can yield up to 1500 kg of dry papain per hectare. Each fruit weighs 1.5 to 2.0 kg with a TSS of 13° brix.

CO 6 (1986)

It is a selection from 'Giant' papaya. The variety is dioecious, dwarf, suitable for papain extraction (7.5 to 8.0 g of dry papain / fruit) and for table purpose. It yields 80-100 fruits per tree. The fruits are medium sized weighing 1.8 to 2 kg, yellow fleshed with a TSS of 13.6° brix.

CO 7 (1997)

This variety is gynodioecious and developed through multiple crosses involving CP 75 (Pusa Delicious x CO 3) and Coorg Honey Dew. It is superior to CO 3 in terms of fruit size and quality. It produces about 110 fruits / tree weighing around 1.0 to 1.2 kg each. The fruits are oblong, attractive, firm, red fleshed with a TSS of 16.7%.

CO 8 (2013)

Dioecious, derived through improvement of CO.2, Red pulped, suitable for dessert, papain and processing, Duration-2 Years and yield - 200-230 t /ha

VARIETIES RELEASED FROM IIHR, BANGALORE

Surya

It is a cross between 'Sun Rise Solo' and 'Pink Flesh Sweet'. It is gynodioecious variety developed at IIHR, Bangalore. It provides medium sized fruits each weighing on an average 600- 800 g with red pulp. The fruits are sweet with good shelf-life. It is quite promising for fruit industry.

Coorg Honey Dew

It is a chance seedling of Honey Dew developed at Central Horticultural Experiment Station, Chethalli of IIHR. It is a gynodioecious type (Hermaphrodite and female trees are produced and no male trees). The fruits are oblong. It is a dwarf and high yielder. The flesh is thick with good flavour. The fruits formed from female flowers are almost seedless.

Varieties released from IARI, New Delhi

Pusa Delicious

This is a gynodioecious type developed at IARI, New Delhi. Trees are dwarf, fruits are medium in size with deep orange flesh colour having food flavour.

Pusa Majesty

It is also a gynodioecious type developed at IARI, New Delhi. The fruits are medium in size, round in shape and having good keeping quality.

Pusa Giant

This is a dioecious cultivar developed at IARI, New Delhi and tree bears fruit at one metre height. The plant can withstand strong wind and storm. Fruits have attractive big size each weighing 2.5 to 3.5 kg.

Pusa Dwarf

It is also a dioecious cultivar developed at IARI, New Delhi with dwarf stature. Trees are more precocious in bearing. The trees bears at 30 cm height. Fruits are oval in shape medium in size, the weight ranging from 1 to 2 kg.

OTHER VARIETIES

Punjab Sweet

It is a dioecious type developed at Punjab Agricultural University. Trees start bearing at 75 cm height. Fruits are round to oval with yellow flesh. The weight of the fruit ranges from 1 to 1.5 kg.

Washington

It is a dioecious cultivar with pink petiole. Fruits are medium to large, each fruit weighing about 1 kg. The fruits are sweet with good keeping quality.

Weimanalao

It is a gynodioecious variety. The fruits are round with orange yellow flesh. The flavour is good and the flesh is thick and firm.

Sunrise solo

The variety is named 'solo' because one man can easily consume one fruit. This is gynodioecious variety from Hawaii with small sized fruits, pyriform and yellowish orange pulp and keeping quality is good.

Red lady

It is a Taiwan variety. The fruits are small. The flavour is good and the flesh is red thick and firm.

IMPORTANT CULTURAL PRACTICES

Propagation

- Papaya is normally propagated by seeds.
- 500 g seeds will be required (200 g/ acre) for planting one hectare.
- The seedlings are raised in polythene bags of size 10 x 20 cm with 2 parts of sand and 1 part each of red soil and well decomposed farm yard manure.
- In each bag filled with pot mixture, 5 or 6 seeds should be sown.
- Seeds germinate readily in 10 to 20 days.
- Papaya seeds loose viability very quickly if stored poorly.
- Seedlings will be ready for transplanting when they are 45-60 days old.

Planting

Pits of 45cm x 45cm x 45cm size are dug at about 1.8x1.8m apart either way. This would accommodate 3000 plants per hectare. Due to sex variations, about 40 to 60 per cent of the plants may turn to be male in the case of dioecious varieties. Therefore, in such case 2 to 3 seedlings per hole at 30 cm apart in the pit should be planted, so that when they reach the flowering phase, the unproductive male trees can be removed to keep the population ratio of one male tree for every 15 to 20 female trees. In the case of bisexual varieties, such contingency may not arise. One good seedling per pit may be planted.The best time for planting papaya is the beginning of the South- West monsoon in most parts of India. In south India, June to October and January to March are suitable for planting as the other months are either too hot or rainy.

Square Method

Rectangular Method

Fig. 20. Planting methods in papapya

Thinning

In dioecious varieties (CO.1, CO.2, CO.4, CO.5 and CO.6), in the seedling population, the male and female plants will be in 1:1 ratio. The male plants are not productive. Keeping one male tree for every 20 females for effective pollination, the excess male trees should be removed. Normally male trees flower earlier than female trees and can be easily identified as they have pendulous, hanging and branched stalk. In gynodioecious varieties, stamens can be noted adhering to petals surrounding the ovary. In the gynodioecious varieties (CO.3 and CO.7) in each pit only one vigorously growing hermaphrodite tree should be retained.

Manuring and Fertilization

Papaya is a heavy feeder and adequate manuring of young and mature papaya tree is essential to maintain the growth and vigour of the tree so as to obtain regular high yields. Before planting, for each pit, 10 kg of well decomposed farm yard manure or compost, 20 g Azospirillum and 20 g Phosphobacterium should be applied. Another 20 g in each of Azospirillum and Phosphobacterium should be applied again 6 months after planting.

The following fertilizers may be applied at 2 months interval starting from the 3rd month of planting (to provide 50 g in each of N, P and K) per pit after the removal of unwanted excess male or female plants.

Urea	:	110 g
Super phosphate	:	300 g
Muriate of potash	:	80 g

If feasible, application of the required nutrients can be also done at monthly intervals as the studies taken up at this department revealed that monthly split application improved yield and other parameters compared to bimonthly applications.

Foliar sprays of zinc sulphate 0.5% and boric acid 0.1% at 4th and 8th month after planting stage increase the yield and quality of fruits. This micronutrient spray can be prepaed by dissolving 5 kg of zinc sulphate and 1 kg of boric acid in 100 litres water. Two more sprays can be given at four months interval for the second year crop also. It is necessary to add a few drops of sticker solutions such as Teepol or Sandovit to the spray fluid before taking up the foliar spray.

Fertigation

Application of 10 litres of water per day + 13.5 g urea and 10.5 g of muriate of potash / week through fertigation and soil application of super phosphate 278 g per plant in bimonthly intervals improve growth, yield and quality of fruits in CO.2 papaya.

Irrigation

Papaya responds well to copious irrigation in well drained soils. Regular irrigation helps fruit development and induces the tree to bear larger sized fruits. Water stagnation should be avoided. In most parts of India; papaya are irrigated once in 8 or 10 days.

Intercropping

During the pre-bearing age, short duration vegetables like cabbage, cauliflower, onion, chillies, radish, etc. can be grown as intercrops. Weeding should be done regularly to keep the field weed free in the young plantation as in the grown up field, the interspace remain well covered with the top growth which helps in checking weeds.

Sex expression

Many sex forms such as dioecious, hermaphrodite, gynodioecious etc., have been reported in papaya. There, are no distinct or definite methods to ascertain the sex of the plants at the early stage itself. Besides, many factors have been reported to influence the sex expression.

1. **Environment**: Low temperature tends to produce perfect flowers on the male tree and female flower production is increased in cool weather and short days. Season of planting also affects the sex expression. Planting during February shows more male plants while planting in March/April produces an equal number of staminate and pistillate plants.

2. **Use Growth regulators:** such as GA_3 (50 ppm), ethrel (200ppm) SADH (250ppm) and phosphon - D (2500ppm) increase the femaleness in dioecious types. After flowering the fruits mature within 5 months after flowering; Under adverse climatic condition flowers & fruit drop is very common, to avoid this problem spraying of planofix @ 1ml/liter. Fruit thinning is suggested to get good size and quality.

S.No.	PGRs	Effect
1	GA_3 25 ppm.	Early flowering
2	GA_3 25 ppm 45 and 75 days after planting.	Sex expression
3	GA_3 50 ppm 45 and 75 days after planting and ethrel at 200ppm.	Increased papain yield
4	Increased yield	25 mg paclobutrazol + 0.4 % amino acids along with 300 g N.
5	200 ppm ethrel or ethephon.	Advancing maturity

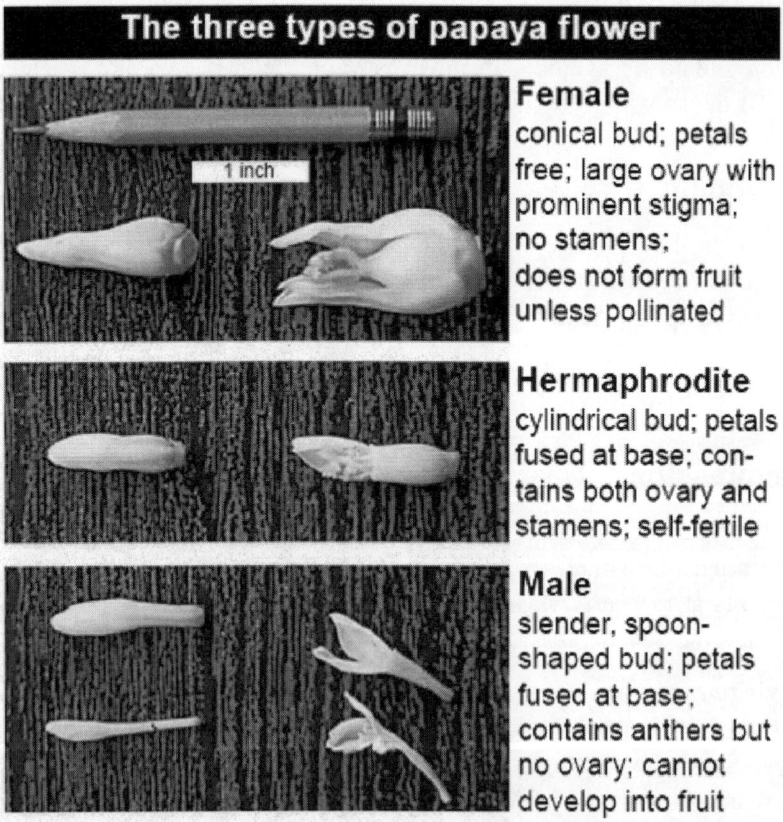

Fig. 21. Different types of papaya flowers

Male tree **Female tree** **Hermaphrodite tree**

Fig. 22. Different sex forms in papaya (Male, Female and Hermaphrodite)

Extraction of Papain

Fig. 23. Papain extraction

The latex or milky juice of the unripe green papaya fruit contains a large amount of digestive enzyme called papain which is able to digest the protein in our feeds. Fully developed green large sized hard papaya fruits which are about three months old are selected for tapping. The latex is obtained by making scratches or shallow incisions on the skin of the fruit The incisions are about 0.3 cm dccp. Usually not more than four incisions per fruit at equal distance are made every day. To cover the whole surface around the fruit not more than five tappings at intervals of four or five days would be necessary. Non-metallic instruments should preferably be used in tapping and collecting, as the juice acts upon metals and gets discoloured. An ivory blade or a sharp edge or piece of bamboo splinter may be used. The latex should be collected in porcelain glass or earthen containers. After about 2 to 4 hours, the latex is scraped out from the tray and dried in the sun. Tapping should be undertaken early in the morning so that drying in the sun can be done before mid-day. This makes the material sufficiently dry by the evening. When thoroughly dried, the latex becomes crisp and flaky. It may be then ground into a powder, preferably still warm. The dried papain is powdered and sieved in 10 mesh sieves. The cream coloured powder should be placed in air-tight bottles or poly bags. Papain can be also dried artificially at temperature of 50 to 55°c which will attain better colour and quality. Potassium metabisulphite (KMS) at 0.5% may be added to it for better colour and keeping quality. The papain production is influenced by certains factors such as fruit size, fruit maturity, varietal factor etc.

IMPORTANT PRODUCTION PROBLEMS IN PAPAYA

1. Fruit Carpellody (Cat Face)

Fig. 24. Cat faced fruits

- This fruit deformity that makes the fruit look like a cat's face occurs on hermaphrodite plants of some genotypes when the plants are exposed to low night temperatures and high moisture and nitrogen levels.
- Under such conditions, the stamens become joined with the ovary and develop into carpel-like structures.
- This deformity makes the fruit unmarketable.
- To avoid this problem, do not use seeds from the affected tree to produce seedlings.

2. Bumpy Fruit

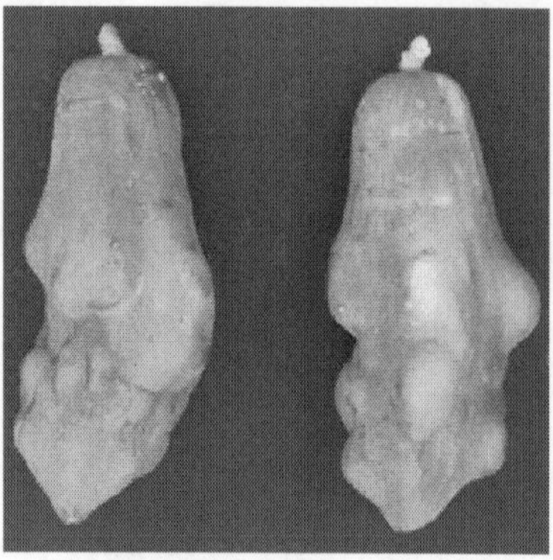

Fig. 25. Bumpy Fruit

- Occasionally, bumpy fruit is observed on the papaya tree.
- This is due to boron deficiency. Latex exudes from parts of the fruit.
- In severe cases, the tree may have a stunted bunchy appearance.
- The seeds are either aborted or poorly developed.
- Spraying the tree with 0.25% borax or a soil application of 0.5-1 gram of borax per tree can alleviate the problem.

3. Leaf yellowing due to water logging

Fig. 26. Yellowing of leaves

- Sometimes leaves of papaya trees turn yellow during the wet season.
- This is a common problem in low lying areas.
- Planting on a mound and improving soil drainage can prevent this problem.

4. Papaya Ringspot Virus

- This viral disease is caused by the papaya ringspot virus –P strain.
- It is transmitted by the insect vectors belonging to the Aphididae, namely *Myzus persicae* and *Aphis gossypii.*
- It can also be transmitted through contaminated pruning or harvesting knives.
- It was first reported in Hawaii and then in Florida, the Caribbean countries and Australia.
- It is present in a few Asian countries, including Thailand, Vietnam, the Philppines, India and Taiwan.

Control measures

- The disease causes mottling and distortion of leaves.
- It produces ringspot symptoms on fruit and water-soaked oily streaks on stems and petioles.
- Fruit production drops and the plants are stunted and in severe cases, the tree dies.
- Since there are no remedial measures for this disease, removal of the infected trees.
- Vector control in the infected areas, as well as sterilizing pruning and harvesting knives, can prevent its spread within the orchard.

a. Plant b. Fruit

Fig. 27. Papaya ring spot virus

5. Phytophthora Collar Rot

Fig. 28. Phytophthora collar rot

- This soil-borne fungus, *Phytophthora palmivora*, infects the collar of the papaya tree trunk at the soil level and the roots.
- The infected collar and roots turn brown and becomes soft.
- In a severe infection, the root and collar will rot.
- Under suitable wet conditions, a white fungus may be seen on the infected collar.
- The older leaves of the infected tree turn yellow and eventually drop, leaving only a few small leaves at the apex.
- The tree wilts and dies.

Control measures

- Planting the papaya seedling in an area not prone to waterlogging or planting on a mound can reduce the incidence of Phytophthora collar rot.
- The fungus can sometimes attack the fruit and removal of fallen infected fruit from the ground helps to remove one of the sources of infection.
- Improving the soil drainage in waterlogged areas can reduce the incidence of the disease.
- If detection of the disease is early, treatment of the collar and roots with the fungicide, propineb, can control the disease.
- However, in severe cases, fungicides are not effective.

6. Fruit Anthracnose

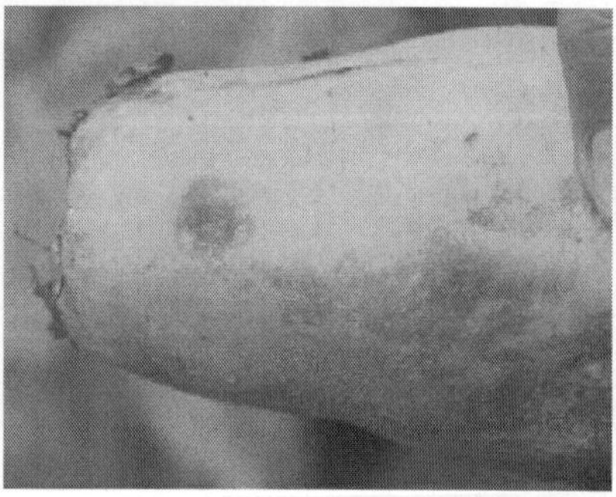

Fig. 29. Anthracnose

- This post-harvest disease occurs when the fruit ripens.
- Small water-soaked spots appear on the surface of the fruit.
- These brown to gray spots are depressed and enlarge rapidly as the fruit ripens to form sunken lesions, a characteristic of the disease.
- Surrounding the spot is a circular translucent margin.
- Sometimes, the spot is covered with masses of light orange to pink conidia.

Control Measures

- To reduce the fungal inoculum in the papaya orchard, do not leave ripe fruit on the tree.
- Practice good farm hygiene by removing fallen fruit from the ground.
- For commercial fruit production, a post-harvest treatment using hot water treatment and chilling can delay the disease occurrence.
- For some export markets, dipping the fruit in a 0.05% a.i. (active ingredient) of the fungicide, prochloraz, is the practice to delay the onset of the disease.

7. Papaya Mealy Bug

Control measures

i. Chemical

- Profenophos or Dimethoate (2ml / lit.) can be combined with neem oil 20ml / lit. or nimbicidine 10000 ppm 2ml / lit.
- Add 1 ml Teepol or Sandavit / lit.
- After fortnight go for the second spray.

ii. Biological

- Many coccinellid beetles and hymenopteran wasp prey / parasitize the insect.
- Lycaenid predator: Spalgis epius larva feeds oraciously on different stages of mealy bug.

Fig. 30. Mealy bug infected papaya fruit

Harvest

Papaya plants produce their first fruits in about 10 months from planting. The fruits can be harvested when they are still hard and green but turn slightly yellow.

Yield

The average yield of different varieties of papaya are as follows.

CO 1	-	100-120 t/ha
CO 2, CO 4, CO 5 & CO 6	-	180-220 t/ha
CO 3	-	100-120 t/ha
CO 7	-	160-180 t/ha

Post-harvest Technology

Fruits are sorted in the field and packed into single-layer boxes with tissue or foam padding to avoid bruising. Fruits can be stored at 30°C and 100% humidity for better color expression prior to packing. To reduce post-harvest fruit rotting papayas are commonly dipped in hot water (48 to 52°C) for 20 minutes, then rinsed in cool water. Fungicides and radiation treatments have also been used for this purpose and to kill fruit fly eggs that may present under the skin. The shelf life at room temperature ranges from 3 to 8 days, depending on storage atmosphere. CA storing is done at 12°C which allows ripe fruit to be stored for up to 12 days. At storage temperatures below 10°C papayas experience chilling injury.

❑❑❑

CHAPTER

12

GUAVA

Botanical name: *Psidium guajava,*

Family: Myrtaceae

Chromosome No: 2n = 22

Origin: Tropical America (from Mexico to Peru)

Guava (Amrud) is one of the most common fruits in India.It ranks 4th in area and production after mango, banana and citrus. It is the fourth most important fruit of India in respect of area and production. It is said to have been introduced from tropical America. It is grown in many parts of the world. It is also popularly called as apple of the tropics and poor man's apple widely naturalized in the country. As a cheap nutritious fruit with a wide adaptability to diverse climatic and soil conditions, the guava undoubtedly deserves greater popularization among the orchardist. Allahabad has the reputation of growing the best guava in the country as well as in the world. It is a very hardy sub-tropical plant, prolific bearer.

Nutritive Value

It is very rich and cheap source of vitamin C (100 to 260 mg per 100 g of the pulp) and contains a fair amount of calcium. It makes an excellent jelly and does not lose the vitamin C in the preserved forms. Guava is the rich source of Vitamin C, and a fair source of Vitamin A. The vitamin C content of Guava is 2-5 times higher than oranges. It also contains Vit-A, B2 and minerals like Calcium, P, acidity 2.4 per cent, carbohydrates 9-10 per cent, TSS 13 per cent, pantothenic acid, riboflavin, thiamin and niacin, also rich source of pectin.

Climate

It is a subtropical and tropical fruit which requires a distinct winter for developing good quality. Guava cultivation can be extended to varying agro-climatic regions owing to wider adaptability. Guava can be successfully cultivated both under tropical and subtropical conditions. It does well up to an altitude of 1,000-1,500 meters. It can thrive in semi arid tracts of India and enjoys cooler climate upto an elevation of 1000 m but it cannot withstand frost. It grows best with an annual rainfall below 100cm restricted between June-September. Places having more than 250cm rainfall are not suitable for guava. Optimum temperature requirement is 23-28°C.

Soil

It is a hardy fruit which can be grown in poor alkaline or poorly drained soils without any manuring or irrigation. It can grow in soils with pH ranging from 4.5 to 7.5 but the best soils are deep, friable and well drained. Since it is a hardy fruit crop, it can be grown on alkaline soils wastelands etc. It is sensitive to waterlogged conditions. If the soils are having a pH of 7.5 and above there are more chances of getting guava wilt. Some varieties like Lucknow- 49 can be grown in saline soils also.

Varieties

Guava varieties are generally named according to the shape, colour or smoothness of skin or from their place of origin.

CLASSIFICATION OF VARIETIES

Classification of Varieties Based on Seed Content

i. Seeded varieties (highly seeded to less seeded types-diploids

ii. Seedless varieties (triploids).

Seeded guavas are more commercial, high yielder's with excellent quality. The seeded guavas are diploids. Seeds range from 250-500 / fruit. Completely seedless and partly seeded, are borne on a plant of seedless variety. The completely seedless fruits develop on the shoots arising from the stem and these are bigger in size and irregular in shape. The partly seeded fruits are borne on normal shoots at the periphery and are small in size and round in shape. Seedless guavas are triploids with irregular shape and low productivity and plants are very vigorous in growth. Hence, unfit for commercial cultivation.

Guava varieties can be mainly grouped into two groups based on the flesh colour.

i. Red fleshed

ii. White fleshed

Among these two, white fleshed are more common and red-fleshed are less common.

Important seeded, seedless and hybrid varieties commercially grown are

Seeded Varieties

1. Allahabad safeda, 2. Lucknow-49, 3. Arka Mridula 4. Red fleshed and 5.Allahabad Surkha etc.

Seedless varieties

Nagapur seedless and Saharanpur seedless.

Hybrids

Saefd jam and Kohir Safeda.

Anakapalli

The fruits are slightly oval in shape with a fruit length of 5.9 - 6 cm. The average fruit weight is 95 g. The seeds are soft and fleshy. It has 393 mg of vitamin C per 100 g of flesh.

Chittidar

It is a white fleshed variety with spherical fruits. Each fruit weighs 95 g. Pulp has good taste and flavour. At calyx end, longitudinal furrows are seen. Skin is thin. The pulp has 108 mg of vitamin C/100 g.

Lucknow-46 (Guja guava)

Fruits have got a short neck at the pedicel end and broad tip resembling the pear fruits (Pyriform). Each fruit weighs 80 g, size 7.4 cm long and 4.9 cm dia. Flesh is white, meaty with good taste. Vitamin C is 130 mg/ 100 g of flesh.

Lucknow – 49 (Sardar guava)

Fruits are round, fairly bigger each weighing 150 – 170 g with a rough surface. The flesh is white with good taste. It has a good adaptability to various localities.

Nagpur Seedless

The fruits are round with thin yellow coloured skin having warty surface. Each fruit weighs on an average 100 g with a size of 4.0 x 4.6 cm. The pulp is white with pleasant aroma and taste and almost devoid of seeds. Vitamin C content is 340 mg / 100 g of flesh.

Saharanpur Seedless

The average fruit weight is 65 g with the size of 4.5 x 4.9 cm. The skin is yellow with white pulp, seeds will be very much limited. The vitamin C content is 250 mg / 100 g of flesh.

Arka Mridula

It is a selection from open pollinated seedlings of the variety Allahabad Safeda developed at IIHR, Bangalore. Fruits are round, smooth with few soft seeds. Flesh is white. The TSS is 12°brix. Flesh is firm and keeping quality is good. It is a heavy yielder.

Hisar Safeda (H25 – 25)

A cross between Allahabad Safeda x Seedless. It has upright growth with a compact crown. Fruits are round weighing 92 g each, pulp creamy white with less seeds which are soft. TSS is 13.4 % and ascorbic acid 185 mg / 100 g. Developed at HAU, Hisar. Yield is 114 kg / tree / year.

Hisar Surkha (H12 – 34)

A cross between 'Apple colour' and 'Banarsi Surkha' pink fleshed hybrid. Fruits are round each weighing 86 g; pulp pink with 13.6%, TSS 0.48% acidity and 169 mg of ascorbic acid / 100 g. It yields 94 kg / tree / year.

Red Fleshed

Fruits are round with 4.5 to 5.0 cm diameter. Each fruit weighs on an average 50 g. The pulp is red with moderate flavour and taste. Fruits are heavily seeded. The flesh contains 385 mg of vitamin C/100g.

Hafsi

Fruits are round with thin skin. The diameter is 6.5 cm and each fruit weighs 142 g. The pulp is red in colour with good taste and flavour. Among the red fleshed varieties, it is comparatively less seeded. It has 108 mg of vitamin – C/100 g.

Saefd Jam

This is a hybrid variety released from Fruit research station, Sangareddy (A.P.). It is a cross between Allahabad Safeda and Kohir. Fruit size is bigger with less seed and vitamin C is more in comparison to their parents.

Kohir Safeda

It is cross between Kohir and Allahabad safeda. Fruit is large in size, slightly acidic. Yield 300 fruits per tree.

Selection-8

- IIHR, Bangalore (from A. Safeda).
- Round and smooth fruits with soft seeds.
- White flesh.
- TSS - 12°B.
- Firm flesh and Keeping quality good.
- A heavy yielder.

Arka Amulya

- IIHR, Bangalore
- *Seedless x A. Safeda*
- Hybrid 16 - *Apple color x A. Safeda*
- Bright red skin.

Lalit

- Released from CISH, Lucknow.
- Fruits 185 g weight, with yellowish skin.
- Pulp light pink in colour.
- Suited for dessert and processing.

Propagation

In India, guava is commonly propagated from seed which germinates in about three weeks. Boiling the seeds for five minutes, soaking them in water for weeks prior to sowing or treating them in strong sulphuric acid for five minutes facilitates their germination. Propagation through this method is not desirable as the seedlings will take more time to come to bearing and seedling trees differ greatly from the mother plants.

Vegetative propagation through layering is therefore recommended. Both air layering and simple layering have been found to be successful. In about 45 days, layers can be separated from the mother plants. These separate layers should be planted in full size pots and they are hardened by gradually exposing them to direct sunlight. Such hardened layers are ready for planting in about six months. Though it is hard to root semi hard wood cuttings, treating with IBA or NAA at 2000 to 5000 ppm root well under mist conditions. In some places, budding techniques using forkert, shield, patch, chip etc have been tried with different success.

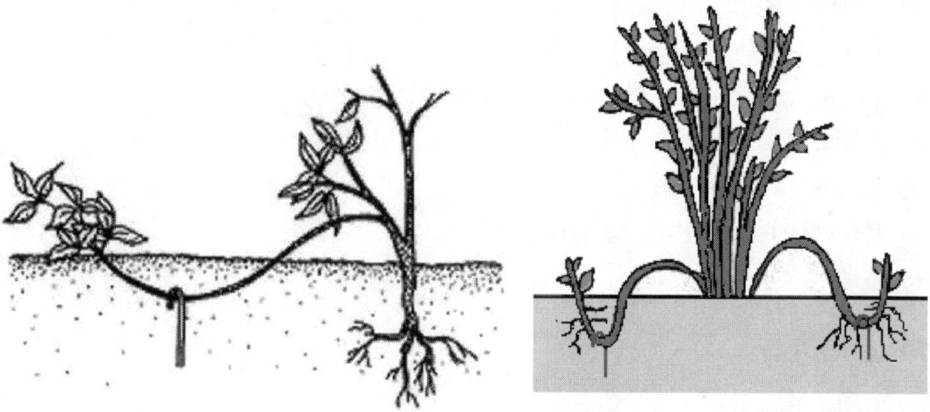

Fig. 31. Ground layering

Planting

The field should be deeply ploughed and properly leveled before planting. Pits size of $75cm^3$ size are dug well in advance to the onset of monsoon. The pits are filled with a mixture of tank silt, FYM Or Compost and soil in equal proportions .To avoid termite problems dusts like lindane are also mixed in the soil mixture. Guava may be planted from June to December in South India and June to August in N.India. Spacing generally adopted is 5-6 meters in square system of planting. In highly fertile soils it may be increased to 6-8meters.

Manures and Fertilizers

A bearing guava tree has to be applied with 50 kg of FYM, 0.5 kg N, 0.5 kg P_2O_5 and 1 kg K_2O per year. This is applied twice during June-July and another during October-November. It responds well to the application of inorganic fertilizers along with organic manures. Therefore for the bearing trees, 50 kg of FYM and one kg in each of N, P and K are applied per tree in two equal split doses, once during March and again during October. The manure and fertilizers are spread in the entire basin of the tree, 15 cm away from the trunk upto leaf drip and incorporated by shallow digging. It also responds to foliar spray of nutrients and spraying of urea 1 % + Zinc 0.5% twice a year during March and October increase the yield.

Guava sometimes suffers from deficiency of micronutrients. Hence, a mixed spray containing $ZnSO_4$, $MgSO_4$. $MnSO_4$ @ 0.5% and $CUSO_4$ and $FeSO_4$ @ 0.25% plus a wetting agent @ 1 ml per 5 litre of solution at various stages viz. new flush, I month after first spray at flowering and at fruit set are recommended.

Irrigation

Guava though can withstand drought, it responds to irrigation at interval of 10 days.

Intercropping

During pre-bearing age, some of the fruits and vegetables can be taken as the remunerative crops i.e., Leguminous vegetables or green manuring crops are grown during first 1-2 years.

Weed Management

Weed is a problem at the early stage of growth, for conservation of moisture, proper utilization of nutrients as well as for effective control of pests and diseases weed free environment is essential. Integrated weed management programme should include growing of cover crops, use of herbicides, inter cropping and hand weeding where ever necessary.

Training and Pruning

a. **Training:** Open centre systems or delayed open centre is generally recommended.

b. **Pruning:** Pruning consists of removal of suckers arising from the base of the trunk. Dried twigs and branches have to be removed and the cut ends may be applied with Bordeaux paste. The flowers are borne on the axils of current season shoots. Light annual pruning after harvesting promotes vegetative growth and flowering. In Tamil Nadu, it is recommended that the tips of 10-12 cm lengths of past seasons shoots are pruned during September and February every year to encourage more laterals. Pruned trees give large fruits and early ripening. When the trees become old, the branches are pollarded leaving 30 cm in length at their origin. The cut branches produce plenty of shoots and flowers and ultimately high yields. In the trees having upright and tall growth habits, the straight growing branches are bent and tied on the pegs driven on the ground. In the bent branches, dormant buds are activated and induced to produce flowers and fruits heavily. In certain parts of Maharashtra, root pruning is practiced to produce heavy yield. In this method roots are exposed and minute roots are cut away and irrigation is withheld so as to allow the leaves to shed. Then, the basins are covered with the manures and soil and irrigated copiously.

Flowering and Fruiting

The guava bears flowers solitary or in cymes of two or three flowers, on the current seasons growth in the axils of leaves. The bearing twigs grow a few centimeters long, putting forth 4-5 pairs of leaves. The blooming period varies from 25-45 days depending on the cultivar, season and region. The initial fruit set is quite high and approximately 80% of flowers set fruits. Afterwards due to severe fruitdrop only 34 - 56 per cent of fruits reach maturity. In cultivars like seedless, the final retention is as low as 6 percent. Under natural conditions, guava tree produces flowers and a fruit twice in a year in North India, but in South and Western India it is thrice i.e. almost throughout the year, which results in no rest period and ultimately guava tree, bears small crops at different times of the year. The pattern of flowering and fruiting periods in guava are:

Special Cultural Operations

1. **Bending:** In medium aged trees, when the trees reach more height, the branches will grow more or less straight up. Such upright growing branches will not produce axillary growth leading to reduction in yield. Such upright branches should be bent and their terminal portions should be buried inside the soil or tied strongly to pegs driven in the soil around the trees. By this, the dormant buds are accelerated and induced to produce new flush which will result in production of flowers leading to higher yield.
2. **Pollarding (or) dehorning:** When the trees become very old, there is practically no production of new flush. In such a situation, pollarding is resorted to. It is nothing but cutting back big branches leaving 30 cm length base at the places of their origin. The new growth in the following season will put forth profuse flowering and give high yields.

3. **Cropping:** The fruit buds are borne on past season growth terminally or laterally. The first flowers are borne on the current season growth in the axils of leaves. The flowers are solitary or in cymes of 2 to 3. The current season growth takes one or two months to bear flowers. The floral buds require 38-42 days for full development. Layers generally take 2-3 years for fruiting. Guava flowers twice a year, first in April-May for rainy season crop and then in August – September for wiriter season crop. In South India, there is a third crop with flowers appearing in October. As the rainy season fruits are insipid and watery and do not keep well.

In certain parts of India, some practices are followed to avoid flowering and fruiting during rainy season so as to get large sized fruits of better quality during winter season. They are:

Bahar Treatment

Consisting of root exposure and or root pruning before the onset of monsoon.In certain parts of Maharashtra, root pruning is practiced to produce heavy yield. In this method roots are exposed and minute roots are cut away and irrigation is withheld so as to allow the leaves to shed. It consists of root exposure and or root pruning before the onset of monsoon. Then, the basins are covered with the manures and soil and irrigated copiously. There are three distinct flowering seasons with corresponding harvesting periods like rainy, winter and spring. It is desirable to take only one crop in a year. In South India the rainy season crop is preferred even though it is of poor quality, since the price is high at this time. In North India, winter crop is of better quality and the fruits also escape the attack of white flies. In Western India, root pruning of guava to regulate the season of harvesting, as is done in the case of mandarins, is recommended in heavy soils only. In lighter soils, withholding of water serves the purpose. This practice is known as bahar treatment.

Bahar	Month			Cropping season	Quality
	Water stress	Flowering	Fruiting		
Ambe bahar	December - January	February - March	July - August	Rainy crop (preferred in South India)	Watery and poor
Mrig bahar	3rd week of April	June-July	November - December	Winter crop	Excellent
Hasta bahar	August - September	October - November	March - April	Spring crop	Good but lowyield

a. **Deblossoming of rainy season crop** - spraying NAA 200-400 ppm and
b. **Withholding of water and removing the soil** from around the upper roots during rainy season and covering it again with soil and manure mixture.
c. **CISH, Lucknow recommends** elimination of rainy season crop by spraying fertilizer grade urea 10-15% twice at interval of 10 days during April-May to get good yield and quality fruits during winter.

Guava fruits should be picked immediately when it is mature and they should not be allowed to ripen in the trees lest the damage by birds and squirrels. Individual hand picking is preferable to shaking the tree. Mature or half ripe fruits are mostly prefered for consumption than ripe or over ripe fruits. Yield varies due to many factors. On an average 800 number of fruits weighing 20-25 kg may be obtained from guava tree.

Harvest and Yield

Guava, being a climacteric fruit, it ripens after harvesting; the fruits are harvested throughout the year (except during May and June) in one or the other region of the country. However, peak harvesting periods in north India are August for rainy season crop, November- December for winter season crop and March April for spring season crop. In the mid climatic conditions of other parts of the country, the peak harvesting periods are not so distinct. change in colour of fruits from dark green to pale green is the indication of maturity, the fruits are harvested at their full yellow but firm for local market, whereas half yellow fruits should be picked for distant markets. The fruits are harvested selectively by hand along with the stalk and leaves.

Yield

Each tree will bear on an average 500-800 number of fruits weighing 20-25 kg may be obtained from guava. The yield per hectare would be 12-15 tonnes/year.

Post-harvest Handling and Storage

Because of their perishable nature, guavas are disposed off immediately after harvesting in the local market and a very small quantity is sent to distant market, they are not kept in cold storage. However, shelf-life of guava can be extended up to 20 days by keeping them at low temperature of 5°C and 75-85% relative humidity. It can also be stored for about 10 days at room temperature (18°C-23°C) in polybags, providing a ventilation of 0.25%.

❑❑❑

CHAPTER

13

SAPOTA

Botanical name: *Manilkhara achras* (*Manilkara zapota/Achras sapota*)

Family: Sapotaceae

Chromosome number: 2n = 26

Origin: Tropical America (Mexico)

Sapota is a delicious fruit introduced from tropical America. It is also known as sapodilla or chiku in India. Sapota fruits are recommended to adult patients ailing from tuberculosis (TB) and children from primary complex. This offers high economic returns even under marginal land and low input management. Sapota is mainly cultivated for its fruits in India while in other countries; it is mainly cultivated for the milky latex produced from the bark and fruits, which yields a commercial product called "Gutta parcha" which is the base for the manufacture of chewing gum (Chicle). Sapota is a native of tropical America and is believed to have originated in south Mexico or Central America and from there, it spread to India and other countries where it adopted very well. Sapota, when fully ripe, is delicious and is eaten as dessert fruit. The pulp is sweet and melting. The sapota fruits are good source of sugar which ranges between 12 and 14 per cent.

Nutritive Value

Sapota is good source of sugars, protein, fat, fiber and minerals (Ca,P,Fe).Sapota is a delicious dessert fruit. The latex from stems and immature fruits is used in the preparation of chewing gum. Fruits can be dried and made into nutritious powder which can be used in Milk shakes and sweets. In countries (Indonesia) Young leafy shoots are used in salads or as vegetable. Sapota is supposed to be medicinal also seeds as diuretic, bark as tonic, antipyretic, febrifuge and in curing biliousness and febrile attacks. sapota is a very sweet fruit when ripe. It contains 15-20% sugars. Sapota pulp is extensively used for making sweets and halwas.

It is also an ingredient of fruit salads and milkshakes. The milky latex secreted by unripe sapota fruits, known as chickle, forms the base for making chicklet and chewing gum. Chickle has also got good market potential. Another bitter principle found in unripe sapota fruits called "sapotonin" has pharmaceutical uses. This way sapota is an easily available low cost fruit with a delicious honey like flavour which is relished by one and all. Its composition is given in table.

Climate and Soil Requirement

Sapota being a tropical fruit crop can be grown from sea level upto 1200 M. It prefers a warm and moist weather and grows in both dry and humid areas. Coastal climate is the best suited. Areas with an annual rainfall of 1250-2500 mm are highly suitable. The optimum temperature is between 11°C and 34°C at higher temperature above 43°C during summer, the flower and fruitlets may drop. Sapota being a hardy tree can be grown on a wide range of soils with good drainage. It come up well in alluvial soils of the river banks, sandy loams near coastal areas, red laterite soils of the heavy rainfall area and medium black soils. Soil should be well drained without any hard pan. Deep and porous soils are preferred. The most ideal soils are deep alluvium, sandy loams, red laterites and medium black soil. It can tolerate the presence of salts in the soil and irrigation water to certain extent.

Varieties

Cultivars classification based on growth habit

i. Trees with erect growing habit.

ii. Trees with drooping habit.

iii. Trees with spreading habit.

iv. Trees with spreading and inferior quality fruits.

i. Trees with erect growing habit: Branches are appearing in whorls, leafy dark green, broad and oval, fruits large and superior.

ii. Trees with drooping habit: Branches in whorls, leaves light green narrow and elliptical, fruit small, inferior.

iii. Trees with spreading habit: Branches irregular, leaves dark green, broad and oval, fruits medium to large, superior.

iv. Trees with spreading habit: Inferior quality fruits.

Cultivars classification Based on fruit shape

i. Round.

ii. Oval.

VARIETAL IDENTIFICATION

Cricket Ball

It is also known as Calcutta Large. The fruits are large and round in shape. Pulp is gritty, granular and moderately sweet. The variety performs well in both arid and humid climate.

Kalipatti

It is a leading variety of Maharashtra, Gujarat and North Karnataka. Fruits are oval shaped, less seeded with a sweet mellow flesh of excellent quality and mild flavour. Fruits appear singly.

CO 1

It is a hybrid developed by crossing Cricket Ball x Oval at Tamil Nadu Agricultural University, Coimbatore. This variety is superior to either of the parents. The fruits are long oval (egg shaped), medium in size with a mean fruit weight of 125 g. The flesh is granular in texture and reddish brown in colour, the taste being very sweet with a TSS of 18° brix.

CO 2

A clonal selection from Baramasi developed at Tamil Nadu Agricultural University, Coimbatore. The tree is vigorous, conical, fruits are obovate to round with mean fruit weight ranging from 112 – 168 g. The flesh is soft, juicy having a TSS of 23° brix.

CO 3

It is a hybrid derived from Cricket ball x Vavi Valasa. It registered an yield increase of 64.3% and 61.83% over female parent (Cricket ball) and male parent (Vavi Valasa) respectively. Fruits are rich in TSS content (24.2° Brix) and sugars (Total sugars 10.8% and reducing sugars 8.2%). The single fruit weight is 132.5 g and the tree yields 1184 fruits weighing 156.94 kg.

PKM 1

A clonal selection from the variety Guthi developed at Horticultural College Periyakulam of Tamil Nadu Agricultural University. The tree is dwarf in stature. Fruits are of two shapes viz., round and oval. The skin is very thin and the pulp has a buttery consistency, very sweet with a TSS of 24° brix. High yielder with medium size fruits, individual fruit weight ranging from 88 to 120 g.

PKM 2

It is a hybrid between Guthi and Kirthi Barthi developed at Horticultural College and Research Institute, Periyakulam of Tamil Nadu Agricultural University. A high yielder with an yield performance of 1500 to 2000 fruits per tree per year weighing 80 to 100 kg. Fruits are bigger in size and oblong to oval shaped. The average fruit weight is 95 g. TSS ranges from 25 to 27° brix.

PKM 3

It is a hybrid between Guthi and Cricket ball. It has vertical growth habit and hence lends itself for high density planting. Trees bear big sized fruits with oval shape and have cluster bearing habit. The fruit yield is 14 tonnes per hectare.

PKM (Sa) 4

It is a clonal selection from PKM 1 developed at Horticultural College and Research Institute, Periyakulam, high yielder (100.4 kg / tree / year) 138.29% increase over PKM 1. Fruits are spindle shaped suitable for dry flakes production. Pulp attractive with light pinkish honey brown colour, crisp and sweet flesh TSS 24 – 25° brix; less seeded (2 – 3 seeds / fruit).

PKM 5

It is a clonal selection from Virudhunagar developed at Horticultural College and Research Institute, Periyakulam. An offseason bearer, putforth flowers during May and Crops during September.

DHS 1

A hybrid between Kalipatti and Cricket Ball. Tree is vigorous, bearing round to slightly oblong fruits with high yield. The fruits are very sweet having a soft, granular and mellowing flesh with a TSS of 26° brix. The colour of the pulp is light orange. The mean fruit weight is 150 g.

DHS 2

A hybrid between Kalipatti and Cricket Ball. Tree is vigorous, bearing round fruits, sweet, TSS 23° brix. The mean fruit weight is 180 g.

Pala

Pala is popular variety in Andhra Pradesh and Tamilnadu. The fruits are small to medium sized and oval or egg shaped, with apex broadly pointed and are very sweet. The bearing is heavy and fruits are borne in cluster. The fruit has thin skin and good flavour.

Kirthi Barthi

The fruits are small to medium sized and oval shaped. There are 4 – 6 ridges on the rind. Fruit skin is rough, medium thick and buff coloured. Pulp is sweet. Fruit apex is rounded. The fruits are suitable for long distance transport.

Baramasi

It is a popular variety in West Bengal, Bihar and Uttar Pradesh. The fruits are medium in size and roundish.

Dwarapudi

The fruits resemble those of Cricket Ball, but smaller in size. It is popular in Andhra Pradesh. Fruits have a sweet pulp and are in great demand.

Propagation

At present, sapota is commercially propagated by vegetative methods such as air layering or gootee layering, grafting and budding. Air layering is a popular method of propagating sapota in western India, while in other parts of India grafting and budding are practised. Recently attempts have been made to multiply this plant through tissue culture also. The Sapota is not usually propagated from seed as the plant raised from seed grows very slowly, takes longer time to fruit and does not inherit the true characteristics of the parent tree. It is therefore, commercially propagated by vegetative methods such as air - layering and inarch - grafting. Air layering is a popular method of propagation in Wes India. In other parts of India, inarch-grafting is generally practiced.

Inarching or approach grafting is the commercial method of propagation in sapota. Inarching should be done during the growing period when the tree is in active sap flow or active growth, various rootstocks, such as sapota seedling, khirni (Pala or rayan) and mahua seedlings are used. Of all the rootstocks, Pala (*Manilkara hexandra*) has been found to be the most suitable. It is the most compatible rootstock and grafts successful and good quality. However, poor germination of Khirni seeds is limiting its rapid multiplication. Instead of sowing the seeds in pot-mixture or red earth or FYM, they have to be sown in pure sand under partial shade. By this about 76 per cent of the seeds will be able to germinate in 43 days. The rootstocks are raised in tube pots. The scion remains attached to the parent tree till the union is complete and if the scion branches are high; the stock plants are placed on bamboo platforms or any such other device.

Name of the rootstocks	Demerits
Mahua (*Madhuca latifolia*)	Rootstock tends to overgrow scion, causes it to die back, fruits are also poor in quality.
Mee tree (*Bassia longifolia*)	Exhibits incompatibility by over growing the scion, pronounced distortion at the graft union and the grafts die prematurely.
Sapota seedlings	Grafts are slow in growth
Manilkara elengi	Grafts not successful

Grafting Time/Period

- Approach grafting during: February (Jan-March).
- Soft wood grafting during: May-July.
- Air layering using 10,000 ppm (1BA+NAA) gives good rooting.

Planting

Pits of 1 m^3 are dug and filled with top soil mixed with 10 kg of FYM, 1 kg of neem cake and 100 g of lindane 1.3%. Grafts are planted in the centre of the pit with ball of earth intact. The graft joint must be at least 15 cm above the ground level. The plants are staked to avoid bending or damage of graft joint.

Spacing

8 x 8m. High density planting 8 x 4m.

High Density Planting (HDP)

- Normal spacing - 8 x 8m (156 trees/ha) - productivity is low.
- HDP - to increase productivity
- HDP trial in kalipatti variety - 5 x 5m - highest yield.
- Identification of dwarf rootstock and managemant - HDP.
- Yield of sapota Cv. Kalipatti (6 years)
- HDP (5 x 5m) - 13.6 t/ha.

- Normal (10 x10m) - 5.05 t/ha.
- Yield of PKM 1 (8 x4m) - 18.00 t/ha (HC & RI, PKM).

Manuring

Sapota is rarely fertilized but the following fertilizers are recommended for maximizing the yield.

Age	N (kg/ha)	P_2O_5 (kg/ha)	K_2O (kg/ha	FYM (kg/tree)
I year	30	30	50	10
Annual increase	30	30	50	10
For bearing trees	150	150	250	50

Manures and fertilizers may be applied in September - October, 45 cm away from the trunk upto the leaf drip and incorporated. Manuring should be done in 2 spilt doses coinciding with the monsoons.

Bearing Habit

- Flowering of Sapota occurs in two seasons, July – November and February – March
- Flowers are borne in past season shoot
- Flowers are protogynous in nature
- Since it is a cross pollinated crop, pollination is effected by wind
- Application of NAA at 300 ppm at flowering stage and pea stage enhances fruit retention.

Irrigation

Irrigate copiously immediately after planting and on the third day and once in 10 days after words till the graft establishes. Sapota can tolerate drought to some extent but irrigations help in improving the yield. Regular irrigations should be given from planting till the plant establishes well. Later irrigations may be according to need and soil and weather conditions. Insufficient irrigations result in dropping of flowers. Drip irrigation can be more useful. Hardy tree - can be grown in black cotton soils - rainfed condition. Varietal evaluation at RRS, Aruppukottai - PKM 1 & CO 2 - rainfed condition. Excess irrigation - flower, fruit shedding and misshapen fruits. Watering - 6 - 12 days in winter & regularly in summer.Drip irrigation - 20 - 30 l/day/tree – TNAU.

Intercropping

Sapota has long pre - bearing age and till the trees cover the entire area with their canopy, intercrops such as legumes and short duration vegetables or fruit crops like papaya and banana can be grown.

Training and Pruning

All the growths that appear on the rootstock below the graft or bud joint must be removed. After 3 – 4 years from planting, the lower most branches upto a height of 60 – 90 cm may be remove as with an advance in age, lowermost branches, if left undisturbed may weigh down to ground and become unfruitful. Similarly over shaded and crowded branches are also removed periodically to permit adequate sunlight and air circulation.

Recently, it has been established that center opening of the grownup trees above 3 – 4 tires of scaffold branches is beneficial to get higher yield with quality fruits. Similarly, pruning of side branches by retaining 1/2 – 2/3 length encourages new growth, flowering and yield.

Flowering and Fruiting

Sapota starts bearing small crops from the second or third year of planting but economical yield can be obtained from sixth or seventh year onwards. Under south Indian conditions, flowers appear almost throughout the year in several flushes at shorter intervals out of which two distinct season of profuse flowering viz July to November and again February to March is generally observed. Fruit setting is a problem in an isolated tree as self- unfruitfulness and **protogynous** condition exist and the pollination is effected by wind.

Use of Growth Regulators

Overcome by foliar spraying of growth regulators. Highest fruit set and retention in sapota Cv. Cricket Ball - 100 ppm SADH before anthesis and 1 month after flower opening. CCC@4000ppm at FBD (Fruit Bud Differentiation) - increase the number of flowers/shoot. Pruning and Paclobutrazol in Cv. Kalipatti at 5 x 5 spacing - beneficial up to 13 years. Mixed planting of different cultivars is recommended. NAA (100-300 ppm) during flowering and again 15 day's intervals increases the fruit set and the yield.

Harvest

Climacteric fruit - judging of maturity is difficult. It takes seven months from fruit set to maturity. Maturity can be judged by certain external symptoms.

Maturity index

- Fruits at full maturity develop a dull orange or potato colour.
- On scratching, the fruit shows a yellow streak in a mature fruit.
- Brown scaly material disappears from the fruit surface.
- The milky latex is reduced in mature fruit.
- The dried spine like stigma at the tip of the fruit falls or drops off easily when touched.
- The mature fruits are harvested with the stalk intact, by giving a twist.

Yield

It depends upon several factors on an average, 12-25 tonnes of fruits/ha /year can be harvested. Yield start from 3rd year of planting. 7 year old tree - 700 fruits/tree/year. 10 year old tree - 1000 fruits/tree/year.

Ripening and storage

Fruits ripen after harvest in about 4-13 days depending upon cultivar. Ripening can be hastened by treating with ethrel (250-750ppm). The fruits are ripen by keeping the fruits in an air tight chamber with. 5000 ppm Ethrel + 10 g NaOH pellets. Between 12-14°C fruits ripen slowly and keep well for about 5 weeks, ripe fruits can be stored at 2-3°C and 85-90% R.H. for 6 weeks and firm fruits for 8 weeks at 3 to 5°C and 85-95% R.H.

❑❑❑

CHAPTER

14

POMEGRANATE

Botanical name: *Punica granatum*

Family: Punicaceae

Chromosome No.: 2n = 18

Origin: Iran

Pomegranate is a favorite table fruit of tropical and subtropical countries. The edible part of pomegranate is a juicy outgrowth of the seed called the 'aril' contains cool refreshing juice. It is liked for the cool refreshing juice and also valued for its medicinal properties. In India though a number of states cultivate pomegranate, the main state which has the maximum area is Maharashtra. It is called as 'Anar' in Hindi. The tree is deciduous in temperate countries, while it is evergreen in tropical and subtropical regions. The fruits are borne terminally on short spurs, arising from mature shoots. Tree is hardy and bushy having a tendency of developing multi-stems.

Nutritive Value

One pomegranate fruit supplies about 40% of an adult's daily Vit-C requirement. Pomegranate is a rich source of carbohydrate (14.5%), protein (1.6%), calcium (10 mg/100g), phosphorus (70mg/ 100g), iron (0.3 mg/100g) and vitamin C (65mg/ 100g).

Climatic and Soil Requirements

Despite adapted to a wide range of climatic conditions, pomegranate yields the best quality fruits in areas of cool winter and hot dry summer (which is prevalent in Baluchisthan, Afghanistan and Iran). Right from sea level upto 1850 metres, it can be successfully grown. It is fairly tolerant to low temperature, of course with differences among varieties. For proper fruit development and maturity and sweetness, a temperature of 35 - 38°C is needed. Under humid condition the quality gets affected. At higher elevation and areas of low temperature during winter the tree behaves as a deciduous one. It is not very specific about its soil requirement. However, in deep loamy or alluvial soil it gives very good yield. It can tolerate salinity and alkalinity in the soil to certain extent.

Varieties

Because of the hard seeds though the aril is pleasant, consumption of pomegranate has been a tedious and boredom process for centuries. But due to evolution of soft seeded genotypes, there is a great increase in the consumption rate of this fruit.

HARD SEEDED TYPES

Kandhari

Fruits are large with deep red rind, aril deep blood red or deep pink with sweet, slightly acidic juice. Seeds are very hard.

Musker Red

Medium sized fruits with medium thick red rinds. Aril is fleshy with moderately sweet juice, seeds are medium hard.

Alandi or Vadki

It possesses medium sized red fruits, aril fleshy, blood red or deep pink with sweet acidic juice. Seeds are very hard.

Kabul

A large fruit, dark red with yellow parches aril dark red fleshy seeds hard with slightly bitter juice.

SOFT SEEDED TYPES

Dholka

This cultivar has large fruits with greenish white rind, whitish to pinkish white, thick, juicy soft arils. It is the commercial variety of Gujarat.

Paper Shell

Medium sized fruits with thick rind; arils are fleshy, reddish to pink with sweet juice. Seeds are soft.

Spanish Ruby

It has medium sized fruits with thin rind , flesh rose coloured and seeds are soft.

Ganesh

Originally it was identified in Ganesh kind garden, Pune by raising Open pollinated (OP) seedlings of Alandi and designated as GBG-1. Has medium sized fruits, aril is pinkish with sweet juice. Seeds are very soft. Fruit surface smooth, yellow with red tinge, round in shape average fruit weight 325 g, TSS 16.47%, acidity, 0.42% developed at MPKVP, Rahuri, Maharshtra.

Jothi (GKVK-1)

At University of Agricultural Sciencs GKVK, Campus, Bangalore, through an evaluation of mixed OP seedlings of Bessein seedless and Dholka, a promising type GKVK-1 was selected and released. It possesses at attractive yellowish red fruit colour, medium sized fruits, red aril colour and soft seeds. Its yield potentials is 18 tonnes / ha.

Yercaud-1 (YCD-1)

At Horticultural Research Station, Yercaud, Tamil Nadu Agricultural University one superior type (ACC. No. 455) was selected with soft seeds and deep purple aril color. This was found suitable to mid elevation of Shervoroys hills. The fruits are medium in size with easilypeelable rind. Each tree gives 100 – 120 fruits weighing 25 kg. The average fruit weight is 350- 400 g.

CO-1

It is a selection developed at Tamil Nadu Agricultural University, Coimbatore with purple aril and soft seeds.

Miridula

Developed at MPKVP, Rahuri through seedlings selection from an open pollinated F2 population raised after crosses make between Ganesh x Gul-e-Shah Red. Fruit surface smooth, dark red in colour, round in shape. Fruit weight about 250 g juice sweet, TSS 16.32%, acidity 0.47%. Seeds are softer than Ganesh.

Ruby

A multiple cross hybrid developed at IIHR, Bangalore for aril colour and seed mellowness. The hybrid develops dark red arils in winter and dark pink or red aril in summer whereas in Ganesh even though the pink or dark pink aril is developed in winter, it is almost white in summer. Ruby derived certain fruit quality attributes from Ganesh, while a gene for red colour of the aril was incorporated from a Russian variety 'Gulsha Rose Pink'. The fruit skin colour is reddish brown with green streaks. Rind is thin, arils are bold (37.2 g/100 arils), seed soft (2.19 kg/ cm2) each fruit weighs on an average 270 g. Yield is 16 – 18 tonnes/ha.

Amlidana

It is an F1 hybrid (Ganesh x Nana) grows well under tropical climate. With quality fruit attributes Amlidana is superior to sour variety Daru whose trees come up naturally in temperate regions of North India. Its fruits provide more acidic (16.18%) 'anardana' an acidulant commercial product prepared by drying the arils of highly acidic pomegranate which is commercially marketed as condiment in North India for use in culinary preparations which serves the purpose of dried green mango (amachur) and tamarind for souring curry, chutney etc. This hybrid fruits weigh 120 g each with pink bold arils. It yields 56 fruits / tree. Trees are short stature and hence suited for HDP which will give higher fruit yield / unit area.

Propagation

Pomegranate is commercially propagated by rooting of cuttings. Semi hard wood cuttings of one year old are used for rooting. The rooting is improved by application IBA 5000 ppm through quick dip method (10 to 20 seconds). It can also be propagated by air layering or gootee, as well as ground layering.

Planting

Transplanting of rooted cuttings is done during monsoon season. Pits of 60 cm^3 size are dug at a spacing of 4 to 5 M between rows and 2 M within the rows, So that higher yield can be obtained during first 5 years. After 5 years, alternate plant within the row can be removed so as to maintain 4 x 4 M or 5 x 4 M. Before planting, each pit should be filled up with 20 kg of FYM and top soil.

Inoculation of 50 g of phosphobacteria + 150 g of Vesicular Arbuscular Mycorrhizae (VAM) in the roots at the time of planting will help better root growth and establishment.

Fertilizers

	N	P_2O_5	K_2O
After I year	250	125	125
II Year	500	125	125
III Year	500	125	250
IV year and above	625	250	500

Besides this every year 20 kg of FYM should be applied / tree. The manures and fertilizers are mixed and applied in a round basin 1 metre away from the trunk. A week before application of fertilizer, the soil around the root zone is slightly removes and 50 g of phosphobacteria + 150 g of Vesicular Arbuscular Mycorrhizae have to be applied near the feeder roots. Besides farmyard manure each tree is supplied with 10 kg of pressmud, by spreading on the top layer of soil. In sandy soils press muds serves as mulch and prevent moisture loss during summer season. Besides this, press mud also supplies some of the nutrients. Press mud applied plants have come to earlier bearing in sandy soil condition. Application of 375 g N, 187.5 g P2O5 and K2O through fertigation along with irrigation level at 20% wetted area is found best for pomegranate var. Mridula.

Irrigation

Drip irrigation using pitcher pot or tube will keep the soil moisture constant without much fluctuation. This helps the pomegranate to get better establishment in the early stage as well as regular bearing in the fruiting phase.

Training, Pruning and Other Intercultural Operations

Pomegranate is trained as bush. Pomegranate tree has a tendency to throw out lot of suckers. If it is trained on a single stem system and if it is damaged by stem borer then the tree will be lost. Therefore, 3-4 stems are allowed per plant and they are pinched at a height of 1 metre and below the pinched tip of each stem upto 25-30 cm, 2-3 branches well distributed in all direction are encouraged. Such training will help for good maintenance of the tree. In pomegranate the fruits are borne terminally on short spurs produced all along the slow growing mature wood.

They bear fruit for 3-4 years. Every year during winter a light pruning is to be given to shorten the previous season growth so as to encourage fruiting. Besides this, dead and diseased branches, water sprouts (suckers) should be removed periodically. Water sprouts from the base should be nipped at the start of their growth so as to avoid wastage of food material in such growth which exhausts the maximum reserve food. If such diversion of food is allowed then there will be very poor bearing in the trees.

Training Methods

Training of pomegranate plant is important to allow certain number of shoots/ stems per plant. It may be trained as multi-stemmed and single stemmed tree.

Multi-stemmed Tree

This method is preferred in Maharashtra, where in 3-4 stems are retained at a hill and remaining shoots are removed. But yield has not been found to be affected. This will give a bushy frame work to the plant.

Single Stemmed Tree

Train the plants, remove all the side shoots upto 2-3 feet and single stem is left. This operation begins at the time of planting. The main stem is pinched at a height of 1m results in the formation of branches. Only well distributed 4-5 branches on all sides are allowed to grow. Pruning of water shoots, weak crotches dead twigs, old spurs is done regularly. After 10 years, old main stems should be removed by cutting back to make it more productive.

Regulation of Flowering (or) Crops Regulation

To obtain higher fruit yield during a particular period, the pomegranate plants are given a resting period. It is done by withholding of water for about 60 days in advance of the normal flowering; Roots are exposed and is known as bahar treatment. After 2 months, manures and fertilizers are applied and light irrigation is given.

Three to four days later heavier irrigations at normal interval are followed. For this treatment the trees readily respond and produce new growth, bloom and bear a good crop. Flowering is noticed in almost round the year and there are 3 main seasons viz.,

Ambe Bahar

Flowering can be induced in February - March. This is taken in the areas where, enough water is available during hot weather.

Mrig-bahar

Flowering can be induced in June - July, coinciding with the outbreak of monsoon, this treatment is taken in the areas where, water is scarce during the hot weather

Haste Bahar

Flowering can be induced in September - October, where the trees have to be subjected to stress during August - September. This is rather uncertain because of the monsoon rains that occur during this period.

Flowering and Fruiting

The tree starts bearing fruits from 3-4th year and continues for about 25 to 30 years. Economic yield is generally obtained from 6th or 7th year onwards. The fruits are ready for harvest in about 5-7 months after the appearance of blossoms. Fruit cracking is a serious problem. This is mainly due to high temperature coupled with moisture stress at the time of fruit growth and maturity some times it is due to boron and potassium deficiency. The intensity increases if the matured fruits are subjected to drought or heavy rains. Cracking can be controlled by avoiding moisture stress during fruit development, application of recommended dose of 500 g of potash and bimonthly spraying of 0.25% borax and 0.1% urea during the later stages of fruit development.

Harvest and Yield

The fruits are harvested when the skin turns slightly yellow and the fruit gives a metallic sound when tapped. During 4th year, the tree bears 25 - 30 fruits and a 10 year old tree gives 150 - 200 fruits / year. The fruits can b stored for 15 to 20 days under ambient temperature. At 0°C and 80 % RH, they can be stored even for 2 months.

PHYSIOLOGICAL DISORDERS

Fruit Cracking

Fruit cracking is the most serious physiological disorder in pomegranate. In young fruits, this disorder is due to boron deficiency but in fully grown trees it is due to moisture imbalance conditions due to irregular irrigation or rains. A dry spell of drought followed by heavy irrigation or rain results in peel cracking. Fruits of mrig bahar are prone to cracking than the fruits of other season crops (bahar).

Management

Spraying liquid paraffin @ 1% at 15 days interval twice during June. Application of recommended dose of 500g of potash + bimonthly spraying of 0.25% borax + 0.1% urea at fruit development. Spray borax at 0.1 % or calcium hydroxide on leaves and fruits starting after fruit set, maintain soil moisture, spray with GA_3 at 200 ppm in the month of June and grow fruit cracking tolerant cultivars. Cultivars like Bedana Bosec, Khog and Jalore seedless are comparatively cracking tolerant.

❑❑❑

CHAPTER 15

PINEAPPLE

Botanical name: *Ananas comosus (*Syn. *A. sativus*),

Chromosome No.: 2n = 50

Family: Bromeliaceae

Origin: Brazil

The pineapple is said to have been introduced into India as early as 1548. It is one of the commercially important fruit crops of India. It is one of the choicest fruit all over the world because of its pleasant taste and flavour. It is an herbaceous perennial. Throughout the length of the stem it is surrounded thickly with leaves. From the axils of the basal leaves, branches with closely arranged leaves arise, which are capable of reproduction of new plants. These are called 'suckers'. Just below the inflorescence, from the peduncle (fruit stalk) small branches with close stem with densely arranged leaved found above the inflorescence is called 'crown'.

Nutritive Value

Pineapple is a good source of vitamin A and B and fairly rich in vitamin C and minerals like calcium, magnesium, potassium and iron. It is also a source of bromelin, a digestive enzyme.

Climate and Soil

Pineapple adapts well to sub – tropical regions to tropical regions having humid climate (75-78%). It performs well in places receiving an annual rainfall ranging from 100 – 150 cm. it requires an optimum temperature range of 21°C to 23°C. it can be grown upto an elevation of 1100 meters from sea level.

The pineapple plant comes up well in all types of soil except heavy clay soil. It does not tolerate water stagnation and soil should be well – drained. Throughout the tropical regions, pineapple is cultivated in the red or reddish brown laterite soils. The soils contain considerable amount of iron- oxides which help to develop the redcolour. The pH of the soil should be 5.5 to 6.

Varieties

Important varieties which are grown in India are Giant Kew, Queen and Mauritius.

Cultivars or Varieties : Types Queen, Kew and Mauritius

Queen

The plants are dwarf and dense with more number of side slips and suckers, leaves are serrated. Each fruit weights on an average 1 kg. The eyes are deep and the fruits develop golden yellow colour as they ripe. The fruits are more juicy and fleshy, and can be stored for longer periods. But they are not sutiable for canning.

Varieties

Jaldhup and Lakhat.

Kew

The fruits of this group are highly suitable for canning. Fruits are oblong tapering slightly towards the crown. Fruits weight 1.5 - 2.0 kg. Suitable for canning.

Mauritius

Yellow Mauritius fruits are intermediate in size, yellow in colour oblong in shape and fair in taste. Flesh is with more fibre. The crown will be big with thorns. In Red Mauritius the fruits will be broader at base and slightly narrow towards the tip. Fruits are red in colour.

Cayenne: It is a canning variety.

Singapore Spanish: Canning variety

Red Spanish: It is a table variety.

Abacus: Table variety

Propagation

Pineapples can be propagated by almost all the parts except the fruits. Suckers, slips and crowns can be used, but the slips and suckers and found superior as planting materials in comparison to crowns. Pineapple is propagated vegetatively through suckers (which start developing from the axillary buds of the basal leaves) or slips (the sprouts grow from the buds on the inflorescence axil or fruitstalk below the fruits) or crowns (above the fruit). The slips and suckers are found to be superior as planting materials in comparison to crowns. The use of sucker (about 400-500 g weight facilitate the harvesting 15-18 months after planting, while the slips will take 20-22 months. The crowns if used will take 24-25 months for first harvest. Hence it is recommended to go in for use of suckers. At the time of planting, the basal leaf scales should be removed properly and then dipped in 0.025% mehtyl dematon and 0.2% copper oxychloride to prevent mealy bugs and stem rot. Recently the dormant buds in the crowns have been successfully used in tissue culture for rapid multiplication of planting material.

Preparation of Field and Planting

The field should be ploughed to a depth of 30-40 cm to have a fine tilth. The field is laid out into trenches alternating with mounds for planting of pineapple. The depth of the trench should be about 22-30 cm. in hilly areas, the trenches are to be made across the slope.

Single-row System

This method is recommended where labour is scarce and weeds are abundant. This system is very much useful for rations. In this system, rows are set at 60 cm apart and plants are set in rows at 30cm spacing. The main disadvantages are the lack of support afforded to the plants which may topple over especially due to the weight of the fruit. Direct sun rays may also cause 'sunscald' or sunburn on fruit.

Double Row System

Trenches are dug at 90 cm interval and in the mounds two rows are set 60 cm interval and the plants are set at 25-30 cm intervals in the two rows exactly opposite in zig zag manner. Double row planting permits high density planting. Trials conducted in many countries including India reveal that dense planting results in :

1. High yields without affecting the size or quality of the fruit,
2. Overlapping of the basal leaves provides shade and reduces evaporation loss as well as weed growth.
3. The leaves tend to twist and grow upright which provide the fruit a natural covering to prevent sunscald and hence results in uniformly coloured and lustrous fruits.

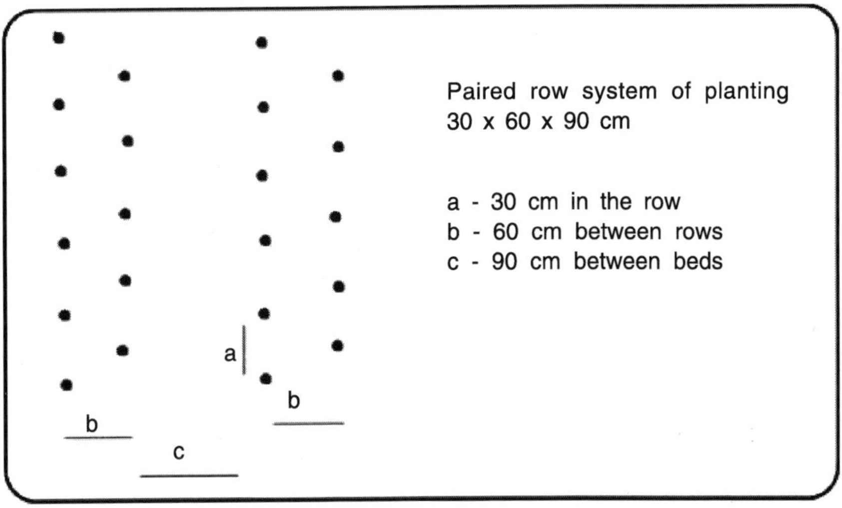

Fig. 32. Paired row system of planting

High Density Planting

High density planting, besides increasing the yield, is associated with other advantages like less weed infestation, protection to fruits from sun-burn and increased production of suckers and slips per unit area, and non-lodging of plants. A plant density of 63,000 plants/ha (22.5 cm x 60 cm x 75 cm) has been found ideal in sub-tropical and mild humid areas of Bangalore. In hot and humid areas of Kerala, Karnataka and West Bengal, a plant density of 53,000 plants/ha (25 cm x 60 cm x 90 cm) performs well. In rainfed, highly fertile and hilly areas, like north-eastern states, a plant density of 43,000 plants/ha (30 cm x 60 cm x 90 cm) has been recommended. The yields recorded under highdensity planting are in the range of 70-105 tonnes/ha. The increase in yield per unit area is in the order of 45-85 tonnes/ha. In nutshell, adoption of high-density planting does not hamper fruit size, quality and canning recovery.

Ratoon

Desuckering should be done immediately after harvest leaving only one sucker on the mother plant. All the slips should be removed. The plants should be sufficiently fertilized and earthing up is done so as to provide good anchorage for the ratoon crop. In high density planting, ratoon can be taken successfully for 2-3 years and in traditional system for 4-5 years. At Hessaraghatta, rate of rationing was compared in traditional and high-density planting of 55,555 plants/ha. In this, 2 successive ratoon crops harvested at 12 monthly intervals amounted to 50.7 and 53.8% of the plant crop yield. The yield of the first and second ratoon crops is to an extent of 50-60% and 40% of the plant-crop in Hawaii and Australia Manures and

Fertilizers FYM 40-50 t/ha. N 16 g, P 4 g and K 12 g/plant in two equal splits at 6th and 12th month after planting. Apply as foliar spray 0.5%-1.0% sulphate of Zinc and Ferrous solutions at 15 days interval to overcome the deficiencies in the early crop phase.

Earthing up

It is a very important operation in pineapple cultivation. This operation gives better anchorage to the plants. Earthing up should be done following each application of fertilizers and also after weeding, hoeing and harvesting crops. This becomes more important for ratoon crop.

Flowering and Crop Regulation

Pineapple plants generally produce flowers after 12 months when they produce about 35-40 leaves and the fruits take 4 ½ - 5 ½ months to ripen. Under normal conditions, only 40-50% of the population comes to flowering.

To have uniform flowering apply the following when the crop attains 35 – 40 leaf stage. NAA 10 ppm + 2 % urea (20 g in 1 lit of water) @ 50 ml/plant poured into crown or 2 % urea + 0.04 % Sodium carbonate + 20 ppm Ethephon (ethrel) @ 50 ml/plant poured into the crown. To increase the size of the fruit, 200 – 300 ppm NAA should be sprayed after fruit formation. To avoid calcium induced Iron chlorosis adequate shade should be given.

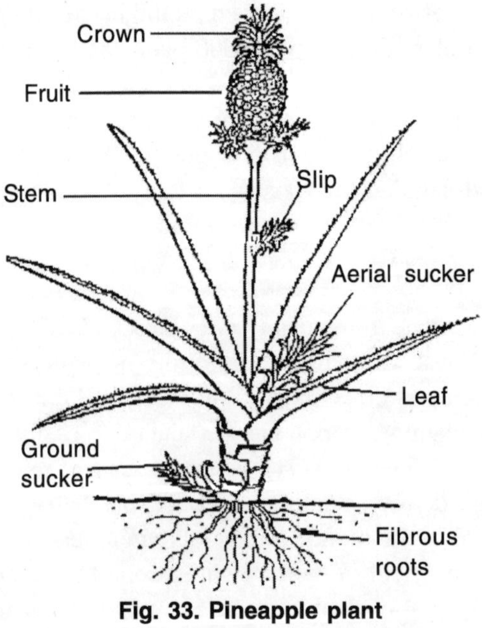

Fig. 33. Pineapple plant

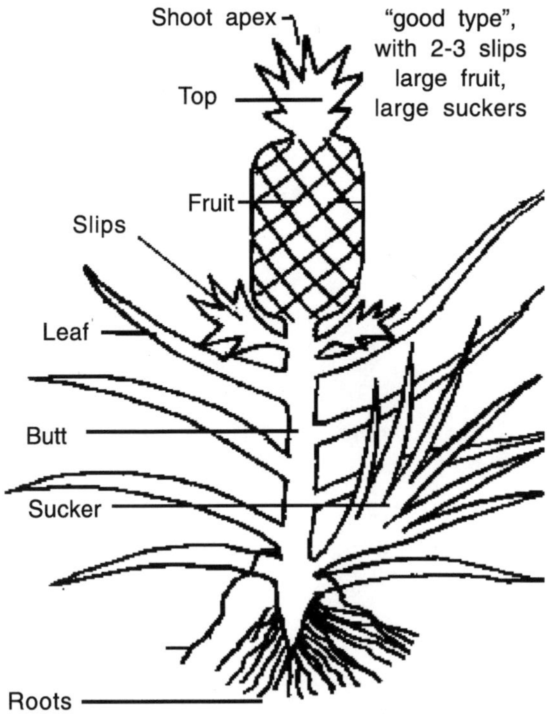

Fig. 34. Pineapple plant parts

Harvesting

Fruits can be harvested from 18 to 24 months. The fruit is ready for harvest when the dark green colour becomes lighter and the deep seated eyes become shallow. Slight colour change at the base of the fruit indicates maturity.

Fruits should be harvested when fully matured for canning purpose and also for distant markets. But for table purpose, the fruits are picked after they develop a golden yellow colour.

Yield

The yield from a plant population of 35000 - 40000 per ha is about 40-50 tones and that from a plant population of 43000-50000 per ha normally varies from 50 to 60 tonnes. Yield of 60-70 t/ha is possible with still higher population of about 63000 per ha.

❑❑❑

CHAPTER

16

JACKFRUIT

Botanical name: *Artocarpus heterophyllus / Artocarpus integrifolia,*

Family: Moraceae

Chromosome No.: 2n = 56

Origin: India

India is considered to be the native home of jack. In the Tamil literature, jack has been given the important second position of significant 'three fruits' viz., 'Mukkani'. Jackfruit is a multipurpose fruit tree. Jackfruit is a common and popular fruit crop of the low land tropics of eastern and southern parts of India. This fruit tree grows well without much care. It is the largest edible fruit (20-40kg), In Srilanka a fruit of about 120 kg was harvested and created the world record. The fruit is also called as **poor man's fruit** in eastern and southern part.

Nutritive Value

The fleshy carpel (which is botanically the perianth) is the edible portion. The ripe fruits are very nutritious. Hundred gram edible portions contains 19.8 g carbohydrate mainly as sugars, fairly large amount of proteins, 1.9 g protein, 0.1 g fat, 1.1 g fibre, 20 mg calcium, 41 mg phosphorus 0.56 mg iron, 175 mg carotene (Vitamin A), 0.03 mg thiamine, 0.13 mg riboflavin, 0.4 mg niacin and 7 mg vitamin C. The seeds contain 7.22% crude protein. One hundred grams of jack carpels supply 88 Kcal of energy.

Climatic and Soil Requirements

Jack grows well and gives food yield in warm humid climate of hill slopes and hot humid climate of plains. From sea level upto an elevation of 1200 m jack can be grown successfully. A temperature range of 22-35° C will be ideal. It can not tolerate frost as well as drought. Under low humidity the bearing is poor and the fruits also do not develop and taste. The West Coast plains with high humidity are found to be very well suited. Proper drainage is ideal for cultivation. The fruit tree is very well suited for dry land conditions. It can be grown under wide range of soils rich deep and well drained soil is good for its cultivation. Though it tolerates a variety of soils, a deep rich alluvial or open textured loamy soil with slightly acidic condition (pH 6.0 - 6.5) and perfect drainage is ideal.

Varieties

Being a cross-pollinated crop and mostly seed propagated, its innumerable types of fruits differ widely in density of spines, rind, bearing, size, shape, quality and period of maturity. Such variations among clones offer great scope for clonal selection.

Jack Types

Cultivated jack types are classified into two broad groups (1) firm flesh (2) soft flesh.

Soft Fleshed

When fully ripe, the fruit yields to the thrust of finger easily. The pulp is very juicy and soft. The taste varies from very sweet-to-sweet and acid to insipid.

Firm Fleshed

The rind doesn't yield to thrust easily .The pulp is firm and crispy. The taste is variable in degree of sweetness. Rudrakshi and Ceylon Jack are the important jack varieties.Rudrakshi is mostly grown for root stock purpose.

VARIETIES

Singapore (or) Ceylon Jack

It was introduced in Tamil Nadu from Sri Lanka. Fruits are medium in size each weighing 7-10 kg. The carpels are crisp, sweet and yellow with strong pleasant aroma. It is a precocious bearer viz., even seedling progenies will start bearing from 3 years after planting (normally in other types the seedlings progenies will start bearing only from 7-8 years after planting). Some trees produce light 'off season' crop between September and December. The fruits are medium in size (7-10 kgs). The flesh is sweet and crispy. The carpels are compacts, yellow and firm with strong aroma, the fruit contain about 80 seeds. Fruits will be available from March - June and again from September to December.

Hybrid Jack

It is a cross between Singapore jack x Veliappala developed at Fruit Research Station, Kallar. Trees are precocious in bearing; carpels are bigger in size sweeter than the parents.

Burliar-1: (T Nagar selection)

It was developed at Fruit Research Station, Burliar. The trees are medium in height and prolific in bearing.

PLR-1: (Palur - 1)

It is a high yielding variety developed at Vegetable Research Station, Palur of Tamil Nadu Agricultural University. A single plant section isolated in Panikkankuppam village near Panruti of South Arcot Vallalar District of Tamil Nadu. The fully ripe fruits have flat stigmatic surface instead of a spiny surface. The special feature of this genotypes is that the trees bear fruits twice in a year viz., fruits will be available in the regular jack season March to June and an off season crop during October to December is also available. Each tree bears about 60-80 fruits. The average fruit weight is 12 kg containing 115-120 flakes. The total flake weight per fruit is 2.36 kg which accounts for 19.68 % of the total fruit weight. Flakes pale yellow in colour, crisp and sweet; TSS is 19o brix.

PPI-1: (Pechiparai - 1)

It was developed at Horticultural Research Station, Pechiparai of Tamil Nadu Agricultural University by clonal selection from Mulagummoodu local. Trees are medium tall maximum bearing in tree truck. On average each tree bears 107 fruits weighing 1818 kg per year in two seasons, viz., April - June and November - December. Carpels are sweet, crisp and tasty with pleasant aroma. It is suitable for commercial planting as well as for planting in home garden.

Propagation

When propagated through seeds, jack exhibits as wide variation among its progenies. Hence vegetative propagation is recommended. Commercially inarching on 10 months old jack seedlings is done to produce vegetative progenies. The grafts come to bearing within 4 ½ -5 years when compared seedlings which take 7-8 years normally. Presence of gum/latex – layering, building, grafting is difficult to carry out. Cuttings were found to root easily under mist by treating with 1BA at 3,000ppm and ferulic acid at 200ppm also used.

Soft wood grafting (cleft method) on 1 ½ months old seedling with scion of 3-4 months old also was found to be successful. Since the viability of seeds is very low, seeds have to be sown immediately after extraction to raise rootstocks. Rudrakshi and *A. hirsute* are also used as rootstocks. In jack modified forkert method of budding is also done. In Malaya air layering after etiolation is reported to be successful. At Pechiparai (Tamil Nadu), patch budding on 3-5 months old rootstock seedlings done during January – February and October-November was found to be very successful. In this method the budded plants are ready for planting in 8 months.

Planting

Jack fruit is planted as a shade tree in coffee garden or as mixed plant, as avenue plant. To have a regular orchard like mango the spacing 10x10 m is followed. Pits of 1 m3 is prepared, 20-30kg FYM, ½ kg SSP and phorate granules (10-20gm) are filled with top soil in pits and planted during June - July. The grafts are planted in the centre of the pits. During summer supplementary irrigation should be provided.

Fertilizers

Fertilizers	1 year after planting	Annual increase(g / plant)	5th year(g / plant)
FYM	10Kg	10Kg	50Kg
N	150g	150g	750g
P	80g	80g	400g
k	100g	100g	500g

The fertilizers are applied during rainy season. If irrigation is available they can be split into two and applied twice in a year June – July and September – October. The manures and fertilizers can be applied in a circular trench taken 50 -06 cm away from the trunk.

Irrigation

Though jack is cultivated under rainfed conditions, it is very sensitive to drought. Hence irrigation should be done depending on the type of soil, season etc. so that, there should not be any moisture stress especially during flowering and fruit set. Similarly too much of soil moisture will affect the quality of fruits. The flakes will 1 develop an insipid taste when there is excess soil moisture.

Intercropping

During the pre bearing age pulses can be raised as intercrop and dried leaves can be spread below the trees to serve as mulch for moisture conservation.

Care of Plants and Weed Management

Jackfruit leaves being a favourite feed for goats, the young plants are frequently damaged by stray goats and cattle unless adequately roots protected by providing gabions for about two years. Gabions may the have to be replaced after a year. Hand watering of young plants during one or two summers is necessary for assured survival and good growth of plants. In colder regions, protection against frost at least during first few years is safe. Cleaning of basins by spading and ploughing of orchards should be followed as routine measures. Frequent weeding and mulching are necessary to achieve normal plant growth.

Training

Jacks are trained to single stem, early side branches should be removed then and there so that a uniform smooth trunk develops for a height of 1.5 – 2 M and then scaffold branches should be permitted to arise. This trunk is going to develop fruit buds and hence should be kept free of growths.

Flowering and Fruit Set

It is a monoecious tree. The male and female flowers of Jack are borne on same plant on separate drooping catkins (Spikes). Male catkins don‘t develop in to fruits. Female spikes are borne on footstalks and also on primary and secondary branches (Cauliflorous), while the male spikes appear both on the foot stocks and on the terminal branches. Flowering starts in December and continues up to March. The flowers are cross pollinated by honey bees. Fruits are multiple types, very large with sharp protuberances, thick endo carp fused with the meso carp, aromatic and rich in latex. The outer part of the perianth is fleshy and edible. It starts flowering from 7-8 years after planting. Flowers start appearing in December and continue up to March. The pericarp around the seed and the fleshy perianth are edible. The fruits take about 90-110days after appearance of the spike, mature in June-July and are available almost throughout the year in Assam & South India. As the trunk increase in size, the active buds of female inflorescence develop from the trunk as well as from main scaffold branches. This type of bearing habit is called ‘cauliflorous’ (Etymology - From Latin *caulis, stem + flôs, flôr, flower.* The production of flowers and fruits on the older branches (or) trunks of woody plants is called cauliflorous). These fruit buds should be protected from any possible damage by rubbing body of the stray cattle such as buffaloes through spreading dried thorny bushes. This is very essential especially when there is no proper fence in the fields. The male inflorescences (catkins) are seen in the current season growth while female catkins are produced as cauliflorous (on trunk and main scaffold). Sometimes there will be more production of female flowers as well as male flowers but there will be no fruitset. This is mainly due to lack of proper pollination. The male catkins show protrusion of anthers on their surface of the spikes. The stigma becomes visible by 8 AM from 4th day after it comes from the sheath. Every day between 8.30 AM and 9.30 AM, the male catkins showing the pollen have to be collected and rubbed on the sticky surface of female of female spikes. This can be done for 10-15 days, for each spike. Such hand pollination will help in proper fruit set and yield. The fruits develop during spring and summer. The fruits become ready for harvest in June or early July. If the flowers on all sides are not pollinated, the fruit does not develop normally. The size is small and the shape may be irregular in less pollinated fruits. Lack of pollination may lead to the drop of female spike too.

Harvest

Normally jack starts producing fruits from 7th - 8th year onwards. Grafted plants can start yield from 4th to 5th year itself. The tree reaches its peak bearing stage within fifteen to sixteen years after planting. In Singapore jack even seedlings start bearing from 3rd year. Normally the fruits will be available from March to June. In higher elevation harvest extends upto September. Even in plains certain genotypes bear an off- season crop during October - December.

Maturity Indices

Dull hollow sound is produced when the fruit is tapped by the finger. The last leaf of the peduncle turns yellow. Fruit spines become well developed and wide spaced. The spines yield to moderate pressure.Development of aromatic odour. Change in fruit colour from green to light yellow. It is a climacteric fruit; harvesting is done by cutting the fruit along with stalk after developing sweet and good flavour. Harvesting is done by cutting off the stalk carrying the fruits For distant market, fruit should be harvested when still firm and without any aroma. The tender jackfruit, which is used for culinary purpose, is harvested before the hardening of the seeds.

Yield

The yield ranges from 20 to 100 fruits/tree. A 15 year old tree yields about 250 fruits, the individual fruit may weigh from 1 kg to about 30 kg.

Storage and Marketing

Jackfruit is not normally stored in cold storage. It has a good storage quality because of its thick peel. A storage life of about 6 weeks is expected when tile temperature is 10°C to 12.7°C and humidity between 85 to 90 per cent. The initial quality and stage of - maturity at harvest are important factors on which the storage life depends. Fruits are stored for 2-3 months at 5°C with 85-90% relative humidity and 1 week under room temperature.

❑❑❑

CHAPTER

17

CUSTARD APPLE (OR) ANNONA

Botanical name: *Annona squamosa*

Family: Annonaceae

Chromosome No.: 2n = 14

Origin: Tropical America (Peru)

The custard apple (*Annona squamosa*) has been growing in India from time immemorial. It is found growing wild in sub-tropical and tropical parts of India. The edilble part of anona is areoles. It has various names like sitaphal, sugar apple and sweep sop. Custard apple can be called as a delicacy of dry region due to its very sweet delicate flesh. It is a deciduous or semi deciduous tall woody shrub of about 5 - 6 meters height having irregularly spreading branches. Tropical America is considered to be the natoive home of all Annonaceous fruits. The probable origin of *Annona squamosa* is West Indies and South America, while *A. cherimola* originated in mountains of Ecuador and Peru.

Relatives of custard apple

Ramphal (or) Bullock's heart (or) Bull's heart: *Annona reticulata:* Fruits are larger in size, heart shaped, smooth and less seeded but pulp is inferior in quality. Used as rootstock.

Lakshmanphal (or) Atemoya: *A. atemoya* (*Annona squamosa* x *A. cherimola*): It is a F1 hybrid with a better quality of *A. cherimola* and adaptability to high temperature as that of *Annona squamosa.*

Hanumanphal (or) Cherimoyer (or) Cherimola (or) Cherimoya: *A. cherimola:* Fruits are the most delicious, slightly acidic and sweet with buttery consistency of pulp and low seed content.

Mullu Ramphal (or) Mamphal (or) Sour sop: *A. muricata:* The fruits have many soft spines. Fruits are highly acidic. Amongst *annona* species it bears the largest fruit.

Other Species

A. *A. glabra* - Pond apple /Monkey apple.

B. The African species *A. senegalensis* - wild soursop.

Nutritive Value

The custard apple fruits are mostly used as table fruits or pulp is mixed with milk or icecream. It is rich in carbohydrates and provides good amount of proteins and minerals. It is a good source of Vit-A and C. Its calorific value ranges from 822 to 1050 K Cal/kg (as compared to 741 K cal/Kg of mango). The fruits are rich in carbohydrates mainly in the form of sugar (23.5%), protein (1.6%), calcium (17mg/100g), phosphorous (47mg/100g) and iron (1.5mg/100g).

Climatic and Soil Requirements

A warm climate (not very hot) with high humidity and mild winter will be the ideal climatic condition for the best growth and yield of most of the *Annona spp.* except *A. cherimola*. Though they can tolerate extremes of temperature (below freezing and upto 40°C), but such extremes won't be ideal for production of flowers and fruits. They can be grown from sea level upto 1000 metres. Where the summer temperature in very high (more than 40°C) and humidity is low there will be no fruitset though the flowering is profuse. An annual rainfall of 500-700 mm is adequate for growth and fruiting. Cherimoyer (*A. cherimola*) requires a lower temperature and a subtropical mild climate and it is very difficult to be cultivated at very high temperature. Atemoyas (the hybrids between custard apple and cherimoyer) have the superior tastes of cherimoyer can tolerate fairly higher temperature just like *Annona squamosa*. Ramphal (*Annona reticulata*) does not tolerate severe summer when compare to sitaphal. Annonas can be grown in varied soil right from heavy clay upto sandy one. They can also be grown on rocky, marginal and even waste lands. However for best yield, a well drained fertile soil with neutral pH will be ideal. The plants are shallow rooted and hence a deep soil is not necessary. They can be grown on slightly alkaline soil and with irrigation water having slight higher pH and salinity.

VARIETIES

Balanagar

It is a cultivar of *Annona squamosa*. The fruits are greenish yellow in colour. Each fruits weighs 130 -140 g. TSS is 20.7^{o} brix.

Red Sitaphal

It is probably originated as chance seedling. Fruits of this cultivar though belongs to *Annona squamosa* are pinkish dark with erythrite red pulp. Average fruit weight is 150 - 160 g with 22.3^{o} brix TSS.

Mammoth

It is a cultivar of *Annona squamosa*. The weight of the fruit is about 125 g; TSS is 24° brix.

Arka Sahan

It is an interspecific hybrid developed at IIHR (Indian Institute of Horticultural Research), Bangalore by crossing *A. atemoya* (Island gen) *x Annona squamosa* (Mammoth). Fruits are big (210 g); skin is light green in colour with waxy bloom, moderately thick with large flat eyes. Fruits have improved shelf life viz., take 7 deys to ripe, 4 days more than 'Mammoth'. The creamy white flesh is juicy with mild pleasant aroma and tender with sparse or less seeds (9 g/100 g of fruit weight). The fruit of this variety is also characterized by large segments or flakes and many of which are seedless. Flesh is very sweet with high TSS (30 o brix) compared to 24° brix in 'Mammoth'. Average yield is 12 tonnes/ha.

APK (Ca) - 1

It is a clonal selection from a high yielding type in State Horticultural Farm, Courtallam of Tirunelveli District of Tamil Nadu developed at Regional Research Station, Aruppukottai. It is a high yielder in rainfed vertisol (Black soil) and yields about 14.90 kg/tree, 30.7 % more than Balanagar. Each fruit weighs 207 g. average number of fruits would be 72/tree. TSS is 24.5^{o} brix, acidity 0.2%. This is adapted to semiarid plains.

Atemoya (*Annona atemoya*) Cultivars

African Pride

It is a cultivar of Atemoya (*Annona atemoya*). It is cross between cherimoya x custard apple. It is a popular variety grown in subtropical region of Australia.

Pink Mammoth

It is a cultuvar of Atemoya (*Annona atemoya*). The pulp is similar to Lakshmanphal or Cherimoyer (*A. cherimola*), being juicy with an excellent acidic flavour.

Propagation

Annona is propagated commonly by seeds. Fresh seeds germinate in 20 - 30 days. Since the seeds loose viability; they should be sown, after removal from ripe fruits. The seeds of cherimoya treated with 10000ppm GA_3 significantly increased germination by 70%. Seed propagation results in variability in plant vigour, prolonged juvenility and inferior fruit quality. Vegetative propagation by budding or inarching on own seedlings and *Annona reticulata* ensures uniformity. Budding is usually done in early spring or in the autumn. Inarching should be carried out in early spring using one year old mature scion and more than one year old root stock.

Planting

A spacing of 5 M x 5 M is recommended for Annonas. Pit size 60 cm^3At planting, 50 g of phosphobacterium + 150 g of VAM (Vescicular Arbuscular Mycorrhiza) are inoculated on the roots and the plants are planted in the pits filled with 10 kg of farm yard manure (FYM) and top soil. This helps in rapid growth of roots and better establishment of plants, especially in dry regions.

Spacing recommended for various *annona species* are:

- Custard apple - 3.0 x 3.0 to 5.0 x 5.0, 5.0 x 7.0m.
- Bullock's heart - 5.0 x 5.0m.
- Cherimoya - 6.0 x 4.0 to 8.0 x 6.0m.
- Sour sop - 4.0 x 4.0 to 8.0 x 8.0m.

Manures and Fertilizers

Since Annonas are mostly cultivated on poor soils, manuring is necessary for production of good crops. Green manure crop like moong and other leguminous crops can be raised. Application of caster cake and bone meal, in 2:1 ratio is found beneficial. Application of 10 kg FYM, 250:125:250 g of NPK in two splits along with 50 kg FYM/Plant/Year. Apply 30 gm each of Azotobacter, VAM, Azospirillum, Phosphobacteria and 50% of 125:65:125 g NPK per tree. The fertilizers should be applied at the commencement of rainy season.

Irrigation

Custard apple is mainly grown as rainfed fruit crop in India. This crop is benefited from the rainfall that is received during the fruiting season. One or two irrigations during fruit development improve the size and quality of fruits.

Training and Pruning

Newly planted seedlings/grafts are supported with wooden sticks. Generally, pruning is not recommended in custard apple. Fruits are borne on old as well as current season growth. Light pruning helps in inducing better branching. Where pruning can be done to avoid over crowding or remove dead/diseased wood.

Flowering and Fruiting

Annona plants starts bearing from 3 years of planting and economic yields starts from 5 - 6 years after planting. Flowering periods extends from spring to the end of the rainy season.

Factors affecting fruit set

- Failure of many flowers to form fruits due to lack of pollination.
- It is desirable to set fruits early in the season, but in India the early flowers produce no pollen and hence natural pollination is difficult. Hand pollination gives about 85 percent fruit set compared to natural pollination (30 percent).
- Dichogamy: The stigmas are not receptive when the pollen is shed and the receptivity also remains for a short period. Some custard apple trees shed pollen on the morning and others in the afternoon.

Use of Growth Regulators

The crop can be sprayed with NAA 20 ppm four times at weekly interval during flowering (March - April) for 3-4 times at 8-10 days interval to encourage fruitset and spraying with GA_3 at 50ppm. Dipping of freshly opened flowers in GA ensures better fruit-set, better fruit retention, increased fruit size and weight with less seed. Irrigation and mulching during summer season also helps to prevent fruit drop.

Intercultivation

Intercrops like groundnut, minor millets, cowpea and linseed can be grown in the initial years of planting.

Harvesting

Custard apple matures/ripens during October-November/December in dry region, whereas in humid areas harvesting is done during August.

Maturity Indices

Custard apple fruits are climacteric, therefore they are harvested when they are mature, firm and plumpy, on maturity, fruit turn light green. The inter aereolar space widens the fruits turn creamy white, the skin between the segments or tubercles turn into light yellow colour. Light green fruit colour, yellowish white colour between the carpels and initiation of cracking of the skin between the carpels may be taken as maturity indices.

Harvest, Yield and Storage

The fruits are to be harvested at correct stage of maturity. The fruits are hand picked. The peak harvest period is October-November. A good bearing tree gives about 100-150 fruits i.e., about 8-10 tonnes/ha. For transporting over distant market, the fruits are to be picked before full ripening and are packed in single layer, well ventilated wooden boxes with soft cushioning material. The fruits are highly perishable and cannot be stored for long duration. It can be stored in good condition up to 7 days by treating them with 8% wax emulsion. The custard apples ripen with in a few days after harvest. The mature fruits can be stored at 15°C to 20°C with RH of 85-90% and low oxygen and ethylene but with 10% CO2. Under such storage conditions, the fruits can be kept intact for 12 - 18 days. Additives like Saccharified starch (1:1); high voltage treatment and packing under nitrogen gas cover, sugar (1:2) were quite effective in extending the storage of pulp at 4°C (45 days) and 18°C (90 days) temperature.

PHYSIOLOGICAL DISORDER

Stone Fruits

The formation of stone fruit may be due to physiological disorder or malnutrition or moisture stress.

- The occurrence of such fruits is common in sithaphal during November - December.
- The Sithaphal plants go to dormancy immediately after the harvest of the crop.
- In neglected trees the dormancy starts very early.

- At this junction the fruits on the tree suddenly turn brown.
- Such fruits are known as stone fruits.
- These fruits continue to stay on the tree even after the fruits have shed or even after new growth commenced in spring.

Control

- NPK (3:10:10) or castor cake and bone meal (2:1).
- Application of super phosphate and bone meal improves crop yield and reduce formation of stone fruits.
- Mulching - Black polythene mulch.
- When trees are cultivated and attended properly fruits reach at harvest earlier than the dormancy period and thus escape development of stone fruits.

❑❑❑

CHAPTER

18

AONLA

Botanical name: *Emblica officinalis* Gaertn. (Syn: *Phyllanthus emblica)*

Family: Euphorbiaceae

Chromosome No.: 2n = 28

Origin: South and Central India

Aonla is in cultivation in India since time immemorial. It finds mention in much ancient literature. It is indigenous to Central and Southern India. It is more popular as a backyard fruit throughout the country. However, commercial orchards can be seen in states like U.P and Gujarat, Aonla fruit is very rich in vitamin C (600 mg /100 mg of pulp, and pectin, therefore, regarded very important for medicinal value in Ayurvedics.

Nutritive Value

Hundred grams of fruit pulp contains 14 g of carbohydrate, 0.5 g protein, 1.2 g iron, 30 mg vitamin B and 600 mg of Vitamin C. Because of the high content of iron (1.2mg/100g) and B vitamin C, probably this fruit extract is used in a number of ayurvedic and homeopathic preparations. The probable centres of origin are the South and Central India, Sri Lanka, Malaysia and South China. Aonla fruit is very rich in vitamin C and pectin, therefore, regarded very important for medicinal value in Ayurvedics. Undoubtedly, it is the richest natural source of vitamin C in India. Tannin, containing garlic acid, ellagic acid and glucose in its molecule, which is naturally present in the fruit prevents or retards the oxidation of vitamin C and renders it as a valuable antiscorbutic (controlling scurvy disease) in the fresh as well as in the dried and processed condition. The fruits are made into preserve (*murabba*), sauce, candy, dried chips, tablets, jellies, pickles, tophies, dried aonla fruits, etc. The aonla powder is superior to synthetic vitamin C in treating deficiencies. Anti scorbutic, diuretic, laxative, antibiotic and anti-dysenteric.

Phyllemblin, obtained from fruit pulp has been found to have mild depressant action on central nervous system. Good liver tonic.

Climate

Though amla is a subtropical fruit, it thrives very well and comes to yield in tropical humid conditions also. If young plants are protected from both extremes of temperature, the mature trees can tolerate right from freezing (0°C) to as high as 46°C. Similarly if the young plants are protected from severe drought during summer and water stagnation during heavy rains, then the adult trees tolerate these two adverse conditions. Heavy frost during winter is not conducive to its cultivation. Warm temperature seems conducive for the initiation of floral buds. Ample humidity is essential for initiation of fruit growth of dormant fruitlets during July– August. Dry spells result in heavy dropping and delay in initiation of fruit growth.

Soil

Since aonla is a hardy plant, it can be successfully grown in variable soil conditions. Though in a well-drained loamy soil the amla trees grow faster, they can grow in heavy clay also provided slight drainage is arranged during heavy rains in their early crop growth phase of first 2 – 3 years. Aonla has great tolerance to salinity, alkalinity and sodicity. About its ability to stand drought conditions, it can be said as if nature has created this plant only for drought conditions. They also fairly tolerate alkalinity in soils (pH 8.5) and irrigation water. The deep root system, reduced foliage, dormancy of fertilized fruitlets (April - June) makes aonla an ideal plant for arid and semi-arid conditions. Aonla can be cultivated in marginal soils, slightly acidic to saline/sodic (pH 6.5-9.5) conditions. Heavy soils or high watertable areas are not suited for its cultivation.

Varieties

Most of the aonla plantations have been raised from seed, therefore, exhibit great variability and the fruits are of inferior quality. There has been no standardization of aonla varieties and they are mostly known on the basis of size, colour. There were traditional three main varieties of aonla namely Banarasi, Chakaiya, Francis. These varieties had their own merits and demerits. Banarasi, an early-maturing aonla, is a shybearing, prone to heavy dropping of fruits with poor shelf- life. Chakaiya fruits are fibrous, smaller in size and also have a tendency to bear heavy crop in alternate years. Francis suffers from severe incidence of fruit necrosis. Considering limitations of these varieties, following varieties were identified at Narendra Deva University of Agricuture and Technology (NDUAT), Faizabad, Uttar Pradesh and released for commercial cultivation. They are Kanchan (NA 4), Krishna (NA 5), Narendra Aonla - 6 (NA 6), Narendra Aonla -7 (NA 7) and Narendra Aonla - 10 (NA 10).

Banarasi

The fruits are large in size flattened oblong with smooth skin, yellowish with characteristic three raised segments. On an average each fruit weighs 38g. It is a shy bearer. Trees are having upright growth habit.

Chakaiya

It is a prolific bearer. The fruits are flat, with smooth skin which is greenish in colour. Fruits are small to medium in size weighing 26 g having a TSS of 10.70 brix. The trees are spreading in nature.

Francis

The branches of this variety are drooping type. Therefore, it is also called *Hathi jhool*. Fruit weight is around 40-45g. Fruits are large (41.5g) with a TSS of 12.00 brix, flattened oval, greenish yellow in colour, soft and almost fibreless. The branches have dropping habit. It is an earlier bearing variety than Banarasi. A 10 year old tree yields about 50 kg fruits while at 20-30 years age it yields 300-400 kg fruits per tree. But in those areas in Uttar Pradesh where calcium or cal- cium carbonate is excess in soil, the fruits suffer from internal fruit necrosis during maturity. This is due to excess of calcium and deficiency of boron.

Varieties from NDUAT, Faizabad

Kanchan: (NA 4)

It is supposed to be a chance seedling of Chakaiya. A profuse bearer with small to mediumsized fruits (32g) flattened oblong; skin is smooth, yellowish in colour Fruit are with higher fibre content, preferred by industry for pulp extraction and manufacture of various products.It has a mid-season maturity (mid Nov.-mid Dec.)

Krishna: (NA 5)

It is a chance seedling selection from Banarasi.The fruits are large (40 g), triangular and conical. Skin is smooth, whitish green to apricot yellow with red spot on exposed portion. Flesh is pinkish green, less fibrous, semitransparent and highly astringent. It is a moderate bearer. It is an early maturing variety (mid Oct.-mid Nov.)

NA 6

It is a seedling selection from Chakaiya, prolific and heavy bearer (10.8 female flowers/branchlet). Fruits are most attractive and shining, medium to large size, flattened and very low fibre content. It is ideal for preserve and candy, owing to low fibre content. It has a mid-season maturity (mid Nov.-mid Dec.)

NA 7

It is a seedling selection from Francis. This is an ideal and promising variety among growers. Fruits are of medium to large size with conical apex. Fibre content is little higher than NA-6. It is precocious, prolific and regular-bearer (9.7 female flowers/branchlet). This is an ideal and promising variety for preparation of products and has a great promise. It is a midseason variety.

NA 10

This is a chance seedling selection from cultivar Banarasi. Fruits are attractive,medium to large in size and flattened round in shape;skin is rough,yellowish green with pink tinge. Flesh is whitish green,Fibre content is higher.It is an early maturity variety.

Variety from ARS, TNAU

BSR-1

It was developed at Agricultural Research Station, Bhavanisagar of Tamil Nadu Agricultural University, Tamil Nadu. The trees are medium in height and spreading and hence suited for accommodating more number of trees per unit area (can be planted at 6 m x 6 m spacing). Fruits are medium in size (27g) with flat stylar end and round calyx end. Each fruit has 6 segments; the juice has a TSS of 18.10 brix. It is high in medicinal properties. Besides, Anand 1, Anand 2 and Anand 3 have been selected as promising strains in Gujarat.

Propagation

Seed propagation which has been in practice has given lot of variation in the progenies. Hence vegetative propagation was resorted to. Soft wood grafting is commercially used nowadays. Budding using 1 year old rootstocks through "T" method (shield) or patch method is successful. *In-situ* budding will be better than budding in nursery and transplanting the budded plants. To raise rootstock seedlings, from the fully ripe mechanically or by drying in sun and the seeds are extracted. The seeds are hard and take long time to germinate. Hence the seeds should be treated with concentrated H2So4 for 3 minutes and then washed with water and soaked in 500 ppm of Gibberellic Acid for 24 hours. Such treated seeds can be sown in nursery bed or polybag filled with pot mixture.

Field Preparation and Planting

After deep and thorough ploughing, pits of 1 cubic metre size can be dug at a spacing of 6 m x 6 m during May - June and filled up with 10-15 kg of well decomposed FYM. Grafts or buddings of aonla are best planted in the beginning of monsoon i.e. in the months of June to July. Since the trees grow to a huge size a distance of 8 to 10m both ways is recommended. In areas with irrigation facilities, planting can be done in spring (February-March). The budded plant or seedlings for *in-situ* budding (especially in dry areas) can be planted at the onset monsoon rains.

Important Note during Planting

Plant at least 3 varieties in a ratio of 2:2:1 for the purpose of pollination and maximum yield. For e.g. in an acre, plant 80 grafts of NA-7, 80 grafts of Krishna and 40 grafts of Kanchan for best results.

Training, Pruning, Intercultivation and Irrigation

In training the trees, initially all the rootstock sprouts should be nipped off then and there. The growth of scion should be trained to develop a low headed one. At about 75 cm to 1 m height the first two main laterals are allowed to grow out. Five to six such laterals with wide angled crotches should be encouraged at fairly well spaced on the trunk so as to have a modified leader fashion. When the crops started bearing early year after the harvest of fruits, dead, diseased, weak and criss-cross branches should be pruned. The water sprouts and rootstock growth should be watched and periodically removed. During early stages of establishment, the plants should be watered periodically especially during summer. Fully mature trees are seldom watered. However, irrigation during April-June one in 15 days will help to encourage fruit set and prevent fruit drop. During early stages of crop growth an inter crops like cowpea, green gram can be grown upto 4-5 years.

Manures and Manuring

At the end of first year 20 kg FYM + 30 g N should be applied. An additional quantity of 30 g of N should be added early year as the tree grows. For a bearing tree, 800 g N, 640 g P_2O_5, 750 g K_2O should be applied along with 30 - 40 kg of FYM every year during rainy season. If irrigation facilities are available, this dose can be split into two equal halves, one half applied during April - May (at fruit set) and irrigated. The other half can be given during September - October.

Orchard Soil Management

It refers to the management of soil in such a way that productivity is maintained and soil losses are minimum. The technique should be compatible to crop behaviour. In aonla, flowering takes place in February and after the fruit set the fruits remain dormant through summer until monsoon when fruits begin to grow and are ready for harvest in December-January depending upon the climate of the place. Therefore, in summer soil should not be disturbed and natural cover can be allowed to grow. In the beginning of the monsoon the basin can be cleared, fertilizer and manurial application can be affected and the place between rows could be brought under green manure to be ploughed in at flowering stage and left as such.

Irrigation

Young plants require watering during summer months at 15 days interval till they are fully established. Watering of bearing plants is advised during summer months at biweekly interval. After the monsoon rains, during October - December about 25-30 litres of water per day per tree through drip irrigation should be given.

Pitcher Irrigation

Water is one of the scare commodities in several areas in India because of its less availability and unsuitability for irrigation purposes. In these localities, often rain water conservation and its recycling through modern methods of irrigation is not feasible because of its exorbitant initial expenditure. Under such circumstances "Pitcher Irrigation" is one of the viable alternative for initial establishment of aonla orchards. The process includes: Dig a pit 30-35 cm away to accommodate a earthen pot of 12-15 litres capacity; make a small hole at the lower side towards the newly planted aonla plant; plug the whole by placing a cloth wick and plaster it with mud; in case soil predominates by clay particles, 10-15 kg sand need to be incorporated for better flow of water to the feeding root zone; place the earthen pot and fill it with good quality of water; Put a lid on container to avoid evaporation; cover the tree basin with paddy straw, banana leaf, sugarcane trash or any mulch; earthen pot is filled as per the need. By this method with 150-200 litres of water per year, an aonla plant can be established particularly in the semi-arid, ravines and salt affected areas of the country.

Mulching

During summer, the crop should be mulched with paddy straw or wheat straw at the base of the tree up to 15-20 cm from the trunk. Mulching with available organic waste/black polythene film and cow dung pasting on tree trunk have been found quite effective in minimizing frost injury.

Intercropping

Aonla is a quite fast growing tree. Therefore, it is better to follow clean culture in aonla plantation except during the rains when leguminous crops such as moong, cowpea, urd etc. can be grown upto 6 years of tree age. At this age plants usually remain in non-bearing stage and their root system and top growth are not much developed.

Flowering and Fruit Growth

- Flowering - on determinate shoots during March and the blooming period lasts for three weeks.
- Opening at March end
- Male flowers - Born in clusters in leaf axils all over the branchlets, appear first and high under stress.
- Female flowers - Upper end of few branches and appear later. Poor fruit set - due to high percentage of staminate flowers.

Pollination

- Natural by wind (or) honey bees.
- Fruit set - 12 to 18 %.
- By hand pollination - 27%.
- Fertilization - 36 hours within pollination.
- Embryo remains dormant - ovary does not grow for 4 - 4½ months.
- Embryo starts growing by August end and continues till November.
- The initial percentage of fruits set may vary from 12 to 18. The initial poor fruit set is due to low sex ratio and lack of pollination or pollinizer or both.

Fruit set

The initial percentage of fruits set may vary from 12 to 18. The initial poor fruit set is due to low sex ratio and lack of pollination or pollinizer or both. Soon after fertilization which takes place within 36 hours of pollination, the zygote rests for 120 to 130 days and endosperm nucleus may also rest for 70 to 80 days after fertilization. Therefore, fruits do not show any symptoms of growth during summer. This zygote dormancy breaks by the end of July as a result of division in endosperm nucleus. Fruits begin to grow with maxi- mum growth in September and attain maturity by the end of November. During this growth phase, flowers and fruits also undergo abscission which has been divided into three waves:

Fruit Drop

- I drop - 70% flowers drop off - in 3 weeks of flowering.
- Due to unfertilized ovaries & degenerated ovules.
- II drop - Fruit lets - June - September.
- Lack of pollination.
- III drop - Half developed - Aug to Oct.
- Hormonal imbalance.

Maturity indices

Stage of maturity at which fruits are harvested has greatest effect on quality, storage life and marketability. Both, early and late harvesting, results in extremely poor shelf life of fruits. It is, therefore, essential to formulate harvest indices or criteria for harvesting the fruit at the right stage of maturity. The fruits are light green at first, but when they mature, the colour becomes dull greenish yellow, or rarely brick red. Maximum vitamin-C content is observed in mature fruits, while immature fruits are acrid and low in vitamin-C content and minerals. Delayed harvest leads to increased Vitamin-C content. It has been observed that the best time of harvesting aonla fruits is February when the fruits have maximum vitamin-C content. Besides, the mature fruits are hard and unyielding to the touch and so are well suited for bulk harvesting as well as distant transportation and marketing.

Amla tree starts bearing after 2 years of planting.

- The fruits are harvested during February when they become dull greenish yellow from light green.
- The mature fruits are hard and they do not fall at gentle touch and therefore vigorous shaking is required.

- Fruits can also be harvested using long bamboo poles attached with hooks.
- A mature tree of about 10 years will yield 50-70 kg of fruit.
- The average weight of the fruit is 60-70 g and 1 kg contains about 15-20 fruits.
- A well maintained tree yields up to an age of 70 years.

Yield

A budded or grafted aonla tree starts bearing fruits from 3rd year onwards after planting, whereas seedling trees may take 6-8 years for initiation of bearing. The vegetatively propagated plant attains full bearing within 8-10 years and may continue to bear for 60-75 years of age under well managed conditions. As far as yield is concerned, the production varies from cultivar to cultivar. 'Banarasi' is a poor yielder as compared to Chakaia and 'Francis'. Banarasi and Krishna are shy bearer, Francis and NA-6 are average bearer, while NA7, Kanchan and Chakaiya are prolific bearer. After attaining their age, a bearing tree may yield from 100-150 kg/tree/year depending upon cultivar and management which may increase up to 250 kg/tree with good care thus giving an average yield of 15-20 t per hectare (Mitra 1997; Pathak 2003).

Grading

The fruits are harvested manually, grading and sorting should done according to their fruit size. Fruits are graded into three types on the basis of their size. Diseased and injured fruits should be discarded. The large sized fruits are mostly used for preserve and candy; small sized for preparing chavanprash and trifla and the blemished fruits for powder and shampoo making.

Packing

Proper packaging of graded fruits in card board boxes or wooden baskets is done. Bamboo baskets are mostly used for carrying the produce from farm to local market.

Storage

Aonla fruits can be stored for 7-10 days at room temperature, and for 7 - 8 weeks in cold storage at 0 - 5°C temperature with 85-90 % relative humidity. Aonla from Uttar Pradesh is being sold in the market of Calcutta, Bombay, Ahmedabad, Amritsar, etc. where it fetches higher prices.

Storage in Zero Energy Cool Chambers

Storage in zero energy cool chambers can enhance the shelf life of fruits. Zero energy cool chambers are designed to provide lower temperature and higher humidity than ambient conditions but without using any power. Five cultivars were evaluated for their shelf life under zero energy cool chamber and results indicated that cultivars Chakaya, Francis and Kanchan could be stored up to 18 days, while Banarasi and Krishna up to 12 days in zero energy cool chamber.

PHYSIOLOGICAL DISORDERS

Internal Fruit Necrosis

This disorder is characterized by splitting of fruits and browning of innermost part of mesocarpic tissues at the time of endocarp hardening followed by browning of the epicarp resulting into brownish black areas on the fruit surface. Depending upon the severity of the disorder, mesocarp of affected fruit turns black and becomes corky and gummy pockets develop. This is due to the deficiency of boron. Spray of 0.6 % borax thrice in the month of September and October at about 10 to 15 days interval control the malady.

❑❑❑

CHAPTER 19

APPLE

Botanical name: *Malus pumila/Malus domestica* (Syn: *M. communis, Pyrus malus)*

Chromosome No.: 2n = 34

Family: Rosaceae

Origin: Eastern Europe and Western Asia

It is the **king of temperate fruits**. 'An apple a day keeps the doctor away'. This is the old proverb which signifies the nutritive importance of apple in human diet.apple has been under cultivation since time immemorial and today more than 80% of the world's supply is produced in Europe where the major producers are Italy, France, and Germany. Other countries which also produce apple are Hungary, USA, N.C. America, China, Old USSR, India, Spain, Switzerland, Iran and South America.

Nutritive value

Apple is a rich source of easily assimilable carbohydrate (13.4%) and it is also fairly rich in calcium (10mg/100g), phosphorus (14 mg/100g) and potassium (120mg/100g). It supplies vitamin B and C.

Climatic and Soil Requirements

Apple is essentially a temperate crop and grown in temperate region of the world. Under subtropical zone at an altitude of 1600 - 2500 m above MSL also, apple can be grown. Very low temperature during the bud rest favours better crop production. The typical temperate types require 1000 hours of uninterrupted chilling below 7°C for winter rest. After the bud break, during the growth, long day hours with high light intensity, warm days (not hot days) viz., 12 – 15°C and cool nights (not freezing nights) viz., 7 – 8°C is favourable for production of quality fruits in large quantities. The freezing soil temperature can kill the trees. A well drained, slightly acidic (pH 6.5 – 6.7) loamy soil with good depth (45 cm or more) is considered to be ideal for apple culture.

Varieties

In India at Himachal Pradesh, varieties like king of Pippins, McIntosh, Golden Delicious, Red Gold, Starking Delicious, Yellow Newton and Granny Smith are grown. In Jammu and Kashmir, Irish Peach, Cox Orange Pippin, Kerry Pippin, Ambri, Lal Cider, Golden Delicious, Lal Ambri, Red Delicious, Sunehari and Razakwar are important varieties under cultivation. In utter Pradesh and Fanny, Cortland, Early Shanburry, Golden Delicious, McIntosh, Red Delicious and Buckingham are important varieties. In the hills of tropical region where warm winter conditions exist, Parlin's Bearty and Tropical Beauty have been judged as the best performers. They require only less chilling for budbreak and flowering.

Red Delicious

It is a midseason variety. Widely grown cultivar throughout the world. Evolved as a chance seedling in Iowa, USA. Fruits are large, oblong conical with 5 knob like projections at calyx end. Smooth skin covered with red streaks on a plain background. Flesh creamy white, tender, crisp, fine grained, sweet and highly aromatic. A number of strains like StarkingDelicious (Royal Delicious), Ruby Red, Red Spur, Red Chief, Richard Delicious have been developed from this cultivar.

Jonathan

It is also a mid season variety, fruits are round to conical in shape medium to small in size and reddish in colour.

Golden Delicious

It is a late variety with moderate vigour. Fruit medium to large, oblong, skin golden yellow with russetted prominent small dots scattered all over. Flesh creamy white, firm, crisp, sweet with a blend of acidity. It is a good pollinizer for all Delicious group.

McIntosh

It is a mid season variety. Trees are vigorous. Fruits are medium in size, oblate round, skin smooth, shining carmine colour on pale green background. Flesh White, develops red shade after storage, tender, crisp, sweet with a good acid blend, juicy and mild flavoured. But this cultivar is scab susceptible.

Bramely Seedling

It is an English cooking apple ideal for stewing especially baking. Skin deep green, waxy sometimes with a slight orange red blush.

Jona gold (Triploid)

Golden Delicious x Jonathan. Developed in New York. Fruits large with attractive red strips and have good storage life.

New Jona Gold

A virus free clone of Jona Gold at Japan with super grade having high red colour intensity.

Kent

A hybrid between (Diplod) Cox and Jonathan. Late maturing variety; fruit slightly conical with dark red colour on a green background. For this, Golden Delicious and Sparten are good pollinizers.

Chaubattia Anupam

Developed at Chaubattia from a cross Early Shanburry x Red Delicious. Fruits medium in size, skin smooth, thin shining yellow almost entirely flushed are striped red; flesh is firm, crisp, creamy white, juicy, sweet with distinctive aroma.

Shamrock

A cross between irradiated McIndosh x Starkspur strain of Golden Delicious. It is green apple identical in appearance and taste with Granny Smith but matures six weeks earlier

Kodaikanal-1

A clonal selection made from Parlin's Beauty at Horticultural Research Station, Kodaikanal in Tamil Nadu. It is adapted to very well to warm winter conditions prevailing inSouth Indian Hills. Fruits is medium weighing 150g, global in shape with crisp flesh which is moderately juice. Fully rips fruits are attractive yellow in colour with crimson drop.

Propagation

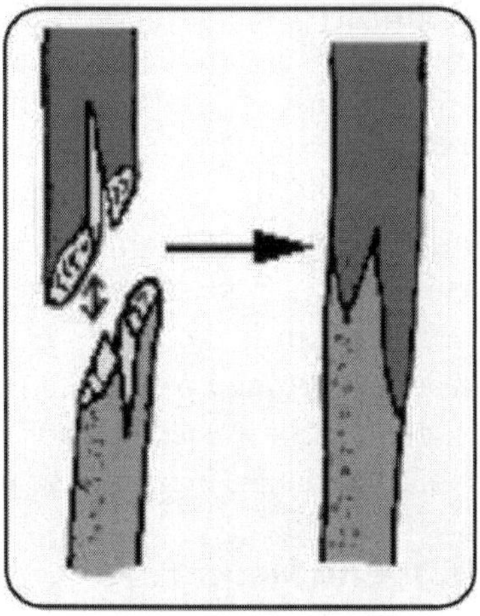

Fig. 35. Whip and tongue grafting

Apple varieties are propagated by whip and tongue method of grafting. The root-stocks are either related species such as *Malus sylvestris* (crap apple), M. *prunifolia and M. sikkimensis* or their hybrid derivatives or seedling progenies of cultivated varieties. The main aims in developing rootstocks are either to dwarf the trees, or to have resistance to below freezing soil temperature or high soil temperature, to wooly aphid, root knot nematodes etc. The East Malling Research Station in England developed Malling Stocks (M). John Innes Horticultural Institute, Jointly Merton and East Malling Research Station jointly developed rootstocks (MM) which, are clonally propagated by either cutting or stooling. Some of the important rootstocks

1. M9 : Dwarfing.
2. M7, MM-106, M 779 & 793 : Tolerant to below freezing (- 40°C to - 35°C) resistant to wooly aphids.
3. Northern spy : Resistant to wooly aphids (*Eriosoma lanigera*).
4. Robusta-5 : Developed at Ottawa, Canada as a selection of *Malus robusta* - Resistant to wooly aphid.
5. M 27 : Ultradwarfing

Planting

Pits of size 60 cm^3 are dug at a spacing of 5 m x 5m. For effective fruit set (since self incompatibility exists in most cultivars) pollinating varieties are interplanted. Usually one in six or one in nine plants is used as pollenizer in self-unfruitful varieties. In intensive system, planting is done at 4 m x 2.5 m or less. A complete row of pollenizer is used for every two rows of main variety.

High Density Planting (HDP) and Medium Density Planting (MDP)

For this dwarfing rootstocks like M9, M4, MM106 and M26 have to be used. The grafts are planted at a close spacing of 2 m x 1m, so that 5000 plants/ha. are accommodated. But in comparison with M. D. P. where in the total number of plants accommodated was 1666 trees/ha (3 m X 2 m), the H. D. P. did not give much yield. An additional yield of 50% more than HDP can be obtained from MDP. The fruit colour and quality also better in MDP than HDP.

Training and Pruning

Training is done to give a desired shape as well as to get a canopy for intercepting maximum sunlight so that all the portions of the tree get sunlight and become productive. The trees are trained as central leader, open centre, modified leader, cordon, dwarf pyramid, espalier, tatura trellis and Lincoln canopy. The apple bear fruit buds in a past season growth called spur. In the established spur system, permanent fruit spurs are obtained on laterals by heading back the leader every year. The strong erect laterals near the central leader are removed leaving wide angled vigorous laterals for formation of spurs. In regulated systems, the crowed weak and shaded branches are removed leaving leader and strong laterals are grow.

Nutrient Management

Apple trees require all the mineral elements. Depending on the soil type, rootstock, a quantity of 500g of N, 250g of P, 750g of K/ tree can be applied.

Irrigation Management

- The drip or trickle irrigation system has high water use efficiency in apples. It also suits the tree in hill slopes.
- Mulching can be done with straw, hay or sawdust and other organic matter.
- Critical stage of water requirement - April to August - after fruit set.
- Summer months – 7 - 10 days interval.
- Drip irrigation - 1,700 litres of water/tree/season - high water use efficiency.
- Conventional system - 3,800 litres/tree/season.
- Requires uniform distribution of rainfall throughout the year.

Weed management

Weed control can be achieved by application or herbicide such as Pitezin (Atrazine + Simazine + Propazine) @ 8 – 10 kg/ha, after weed emergence in 1.5 m band along tree rows. Nearer to root zone littering of straw is advised to maintain organic matter and earth warm activity. Since herbicide can result in a significant reduction of earthworm population.

Floor management

Clean basin management is the common practice of floor management in apple. The basins are kept clean by hand working, hay mulching or black alkathene mulching. In cool climate, black alkathene mulch is very efficient which not only controls weeds and conserves moisture but also improves the fruit quality.

Role of Growth Regulators

When there is chilling deficiency, the bud break will be poor. For bud break, application of Sandolin - A (Dinitro Ortho cresol) at 0.5% in 2% mineral oil emulsion during January is recommended. After bud break, flowering and petal fall. Alar @ 3000 ppm is applies which will encourage the duration of flowering. To prevent fruit drop, AVG (Amino ethoxy Vinyl Glycine) 200 pm is sprayed which can increase the fruit set to 10 fold when applies between pink and petal fall.

Harvest and Yield

- Climacteric fruit.
- Fruits are harvested at fully mature when start developing colour can be harvested.
- Can be harvested based on TSS, ease in separation of fruit from spur, change in ground surface colour from green to pale, fruit firmness and days from full bloom to harvest.
- Hand-picked carefully ad placed softly in the baskets.
- Transferred to packing houses for grading and packing.
- The normal yield ranges from 100 - 150 tonnes/ ha in medium density planting.

Post-harvest Handling

Precooling - Air cooling, cold water sprinkling or fruit washing with water.

Grading - Based on fruit size and quality- 3 grades Extra Fancy, Fancy class I and Fancy Class II

Packaging

- Packed in wooden boxes or CFB boxes.
- 10 or 20 kg fruits perbox.
- CFB boxes – less bruising and fetch good market prices.

Storage

- Long storage life compared to other temperate fruits.
- Cold storage – retards fruit deterioration and decay.
- Storage conditions.

- - 1.0°C to 0°C.
- 85-90 % RH.
- Stored for 4-8 months.

PHYSIOLOGICAL DISORDERS

Bitter pit

This is a physiological disorder, which reduces the fresh market quality of fruit. Deficiency of calcium will lead to physiological disorder in apple fruit called 'bitter pit' or 'cork spot'. Bitter pits appear as slight indentation in the skin usually towards the calyx of the fruit. These areas turn brown and soft dried pits of collapsed tissues develop. In fruits, colour changes and cracks develop. Spraying 0.5 % calcium chloride during the later development of the fruit, 4 times at 14 days interval will help to reduce the deficiency symptom.

Control

Boron deficiency also causes internal browning of fruit and corking spots on the flesh (internal cork) and sometimes as cork tissue in the skin (external cork). Spraying sodium borate @ 1 g/ lit. 3 times during flowering and fruit set will help to correct the disorder. Calcium sprays prior to harvest and calcium dips before storage control the incidence of bitter pit. The plants should be sprayed 45 days prior to harvest followed by a repeat spray after 15 days. The post-harvest dip for 1-2 minutes should be given before storage.

Scald

It is one of the storage disorders in apple. Light mottling on greener surface of fruits are initial symptoms of scald. Darkening becomes more severe with elapsed time and ultimately extends to red surface also. Scald usually affects the skin only but in severe cases it may extend to fruit flesh. The immature fruits are most susceptible to scald which is aggravated by warmer temperatures in storage.

Control

Harvesting at proper maturity and ventilation in cold storage help to reduce the scald incidence. The most common method used to control scald is application of an antioxidant immediately after harvest. Diphenylamine (DPA) is commonly used. Ethoxyquin is also effective for some varieties, but can cause damage to other apple varieties. Antioxidants should be applied within one week of harvest for maximum control.

Sun Burn

This physiological disorder occurs due to intense heat of the sun. Fruit on the southwest side of the tree are generally affected. Water stress can also increase the incidence of sunburn.Initial symptoms are white, tan or yellow patches on the fruits exposed to the sun. With severe skin damage, injured areas of the fruit can turn dark brown before harvest. These areas may become spongy and sunken. Fruit exposed to the sun after harvest can develop severe sunburn.

Control

The best method of control is to avoid sudden exposure of fruit to intense heat and solar radiation. Proper tree training and pruning are critical. Summer pruning must be carefully done to avoid excessive sunburn. Pruned orchards should be regularly irrigated to reduce heat stress. Careful sorting to remove affected fruit upon packing is the only solution once the injury has occurred.

❑❑❑

CHAPTER

20

PEAR, PLUM AND PEACH

PEAR

Botanical name: *Pyrus communis*

Chromosome No.: 2n = 34

Family: Rosaceae

Origin: Western China

Introduction

The probable origin of the Pyrus genus is mountainous china from there it moved east and west and in the different Ecological condition, the evolution would have taken place. Pear is next to apple in temperate fruits. It is a rich source of carbohydrate as sugars, starch and cellulose and minerals like Ca, P and Sulphur.

1. European Pear (*Pyrus communis*)

   Syn. *P. domestica.*

   Tail pears.

2. Oriental pear (*P. pyrifolia*)

   Syn. *P. serotina.*

   Asian types and common pear.

Climatic and Soil Requirements

Pear is adapted to a wide range to climatic condition From as low as 26°C and as high as 45°C. however most of pear cultivars require about 1200 hours below 7°C during winter to complete their chilling requirements in order to flower and fruit to a satisfactory level phase below freezing temperate will affect the crop however after bud break, at blossoming and fruiting severely. Because of its tolerance to wide range of climate and soil, it is grown both in temperate and subtropical conditions. In the tropical belt pear can be grown at an elevation of 1300-2100M above sea level.

Deep, well drained, fertile, medium textured clayey loam soil is the best for pears. A neutral pH range 6.0-7.5 will be ideal. A minimum soil depth 180 cm is required. When compared to apple, pear is less tolerant to drought but more tolerant to wet soils.

Varieties

Pear cultivars can be classified in to 3 groups viz., European types, Asian types and the hybrids of European types and Asian types. The Asian types (oriental pears) are mainly derived from *Pyrus pyrifolia, Pyrus ussuriensis* or their hybrids. The European types are derived from *Pyrus communis.*

European types (Tail pears)

Bartlett (or) William's Pear

It is the most popular commercial cultivar throughout the world (except China & Japan). It requires more chilling hours (1500 hrs) for bud break and flowering. The fruits are ovate pyriform, medium large, green at the time of harvest turning to bright yellow after ripening. Flesh is white, melting, firm and juicy. It has originated in England. High yielding clones like Bartlett Improved-I, Verona –25, Ferrera have been identified.

Anjou

Originated in France. It is fairly resistant to very low temperature and fire blight. Large fruits, skin is bright green when harvest and turning to greenish yellow on ripening. Flesh is fine, mildly acidic. Fruits have high dessert quality and very long keeping quality.

Flemish Beauty

The trees are bigger with more branches. Fruits is large obovate and smooth. The flesh is pure white, very juicy free of grit cells. It is a self-fruitful variety. It can also be used a good pollen source.

Max Red Bartlett

A bud mutant of Bartlett; plants and fruits resemble parent except that the fruit that the fruits colour is dark cranberry red and shoots and leaves have a reddish tinge.

Moonglow and Magnees are two fire blight resistant varieties evolved in USA, Flesh is free from grits cells.

Jorgonelle

It is a European type with superiorquality and adapted to South Indian Hills like Kodaikanal (warm winter conditions).

Starcrimspn

Trees are medium sized fairly upright and spreading. Fruits medium sized, oblong ovate, pyriform dark red change to alternately, juicy, aromatic high in TSS, sweet with excellent eating quality.

Early China

Trees are upright and compact. Fruits round, small, greenish with red blush and very attractive.

Asian Types and Hybrids: (common pears)

Kieffer

It is well adapted to different climatic conditions and moderately resistant to fire blight. The fruit is brownish, gritty and hard. It is a self-unfruitful variety.

Gola

It is found to be suitable for lower altitude. Fruits are large, round and possess excellent keeping quality. Hence it is suitable for long distance transport.

Le Conte

Suitable for lower are mid hills as its chilling requirement is low. Fruits are round in shape, small in size, yellowish green in colour. But it is a blight susceptible variety.

Patharnakh

This is another low chilling variety. Tolerant to very high temperature and hot winds. It possesses a peculiar quality combination of drought tolerance to water logged condition. Fruits are round with prominent dots. Fruits have tough skin and hence suited for long distance transport.

Propagation

Pears are commercially propagated by shield or 'T' budding and also by **whip and tongue grafting**. The rootstock is raised from seeds raised from commercial pear. Number of F1 hybrids of *Pyrus communis* such as Old Home x Farmingdale are multiplied clonally and used as rootstock. Quince is another related genus which shows very good compatibility with a number of commercial cultivars and it is resistant to wooly Aphids, nematodes but susceptible to Oak root fungus, fire blight, cold and excess of lime in soil. Quince clones are easily propagated by semi hard wood and soft Woodcuttings. Some of the commercially used clones of quince are QA, QB, and QC.

But commercial cultivars of pear like pear Bartlett, Bosc, Eldorado need 'Old Home' as they don't have compatibility with quince. The chilled scions of Doyenne du comice grafted at 1.0 m height on 1.0 m long shoots of pear -18 stock found to be the best for higher fruit yield and quality.

Preparation of Field and Planting

One year ahead of planting, the field should be prepared by removing stems and roots of previous trees and shrubs, leveled giving a gentle slope for drainage of Excess water during heavy rains. For a crop on its own rootstock, an initial spacing of 3 M x 2 M is given which is changed to 6 M x 4 M. After 4-5 years. For pear on quince, a planting distance of 3.5 M x 1.1 M is enough since quince has the effect of dwarfing the trees.

The pit size should be 1 M x 1 M x 1 M and the pits filled with a mixture of soil and compost. The planting can be taken up during late fall or early spring. Immediately after planting the basin should be formed and irrigated.

Manures and Manuring

An optimum dose of major nutrients is 600g N, 150g P and 300g K per tree to get the maximum yield. Normally in pears the responses to P and K can be seen only in soils of low availability of P and K .at higher altitudes where soils Ph will be less than 7, the 'P' will not be available.

Similarly, when the soil pH is more than 7 too, the 'P' availability will be less. Under these conditions, application of additional 'P' will increase the yield. N @ 60g /tree in two splits along with a basal dressing of 40g each of p and potash was found the best for Bagugosha cultivar of low chilling pear.

Intercultivation

During summer when young fruits and shoots grow actively, there should be any moisture stress. Sprinkler irrigation will be beneficial not only in the efficient use of water but in application of liquid fertilizer, pesticides and Fungicides too. In order to prevent soil erosion, cover cropping can be done in the interspace.

Mulching can be also done using straw or black polythene to conserve soil Moisture during summer especially in a rainfed crop. Pear trees begin to flower at the age of 5-8 years. Flower bud initiation takes place during summer the spring next year. The flower buds can be increased by application of SADH or paclobutrazol. It suppresses the growth of one year old shoots.

For effective pollination and fruit set in self-fruitful or partially self-Fruitful varieties, across compatible variety should be planted as Very fourth tree in every fourth row which is called pollenizer which will ensure transfer of pollen for effective fruit set.

- Bartlett and Anjou - Self unfruitful
- Beurre Hardy - Self fruitful
- Flemish Beauty, Le Conte and Magness - Self unfruitful
- Other commercial cultivars - Partially self fruitful

Training and Pruning

Pears are trained in a number of systems like pine shaped, pyramid, Spindle, palmette and tatura trellis. Among these, palmette system and tatura trellis are found to commercially superior. In tatura trellis the rows are oriented north – south. Each tree is topped to develop to form 'Y' shape in east-west direction within 50 – 60 crotch angle.

Tensioned wires on steel frames support the arms to a height of 4-5 M and the branches on each arm are trained on these trellises. Bearing trees are pruned by combining heading back and thinning out. Pear bears fruit bud on spurs arising on two year old wood and a spur continues to bear for more than six years.

Thinning

The pear fruits should be done when the crop load is very heavy. Thinning can be done mechanical or chemical method. About 6weeks after blooming 'hand thinning can be done by removing small fruits. Chemical thinning is spraying NAA 25ppm, 3-8 Days after petal fall.

Top Working

At horticultural research station Kodaikanal, Tamil Nadu, top working by which old Trees of country pear are converted in to superior quality types. By this method we can renovate the old country pear orchards through top working with quality varieties. This procedure developed by Nambisan *et al.* is as follows. San pear is grown on an extensive scale in the Kodaikanal and its neighborhoods.

The quality of this fruit is very poor and could not compete with other fruits in the Market. This has rendered pear orchards uneconomic in recent years. To overcome this difficulty, top working the existing pear orchards with superior clones like William, Jorgonelle, Newpear, Kieffer was attempted. For these purpose, trees under 25 years are earmarked. They were headed back about a metre and half from the ground level.

Trunks and there limbs shortened that way were split with the help of a chisel & chamber the lower end scion were be welled in to wedges and inserted in splits of the root stocks. They are then waxed and bandaged with a gunny thread and final with types of polyethylene sheets the graft take ranged from 90- 100% the bandages were removed during July. They came to yield in 3 years it was also found feasible to graft the scion by whip &tongue method on the past season shoot in the crown canopy itself. But it was very laborious and scion should be properly observed for sprout.

The grafting should be done during November to middle of February before the lateral buds should in the stock. By these it is possible to save a lot of time & Old orchards can be converted in to new orchards with superior pear types & the fruits Of such superior types can be got within 3 years of top working.

Harvest, Yield and Storage

Fully mature fruits are harvested while still firm & green for distant market. For local market, they are left on the trees to get better quality fruits. At an interval of 3 - 4 days, two or three pickings are taken up. Fruits should be carefully handled while storage and transit as the bruising is possible by rubbing with one another as well as stalk damage. From a well maintained orchard an yield of 30 - 40 tonnes/ha/year can be expected.

The unripe fruits harvested at optimum maturity can be stored for even 5 months at a temperature of - 1°C. Ripening can be accomplished by keeping at 15 to 21 and 21 to 25°C and 80-85% RH in 3 - 6days depending on the cultivar. Most of the commercial cultivars require this post-harvest chilling treatment for proper ripening. When such post-harvest cold treatments are not available, the fruits can be treated with ethylene, so that they ripen properly and get good quality (both taste and colour).

Physiological Disorder

Premature Ripening

- With pink colouration near the blossom end.
- Brown heart and softening in affected fruit.
- Due to abnormally cool growing season before harvest.
- Night temperature <7.1°C and day temperature < 21°C for few days – cause premature ripening.

PLUM

Botanical name: *Prunus* domestica

Chromosome No.: (Japanese 2n = 16, European plum 2n= 48)

Family: Rosaceae

Family: China

Plum is also one of the important temperate zone fruit crop. They are rich in sugars and carotenes. They are used either as fresh dessert fruit or cooked. Plums are grown on a commercial scale in Jammu and Kashmir, Himachal Pradesh and the hills of Uttar Pradesh and to some extent in the Nilgiris and Pulney hills. Plum fruits supply fair amount of sugar and minerals like potassium, calcium, magnesium, iron, zinc and vitamins like Vitamin C, vitamin A (â carotene) and vitamin B like thiamine, riboflavin, niacin etc. Plums belong to the genus *Prunus* and based on the centre of origin, the following important types are recognized.

S.No.	Common Name	Species	Centre of Origin
1.	European plums	*P. domestica*	Europe
2.	Japanese plum	*P. salicina*	China
3.	Cherry plum	*P. cerasifera*	Westerns and Central Asia
4.	Damson plums	*P. insititia*	Western Asia
5.	American plum	*P. Americana*	North America

Bulk of the cultivated plums belongs to *P. domestica*. The fruit exhibits both yellow and green ground colours and also both red and blue skin colours.

Climate and soil

Plums are grown from sub-tropical plains to temperate high hills because the different species require different climatic conditions. Plums exhibit variations in their climatic requirement and hence it is found growing from the subtropical plains to the temperate high hills of India. European plums demand comparatively more chilling requirement than Japanese plums. In South Indian hills, the plums thrive in hiss station from 1,400 to 2,200 m above MSL.

European plum requires 800 - 1000 hours below 7.2°C for winter chilling for satisfactory bud break. Japanese plums are comparatively lower chilling required group (700 - 800 hours below 7. 2°C). At the same time they are more resistant to high cold than European plum. Since plums bloom early in the season, they are prone to spring frost injury. Generally deep sandy loams with good drainage are the best suited for plum cultivation. European plums perform even better on heavy loam or clayey soil. Red laterite soils with good drainage and rich in organic matter is more suited.

Varieties

Majority of the cultivars belong to European group followed by Japanese plums and Damson plums in order.

European Plums

Green gage, Golden Transparent, Victoria, Stanley

Japanese Plum

Kelsey, Santa Rosa, Starking Delicious, Alubokharo, Rubio, Hale, Gaviota, Beauty, Shiro, Abundance.

Based on the season of maturity, varieties can be grouped as early, mid season and late varieties for South Indian Hills as indicated below.

Early season - (April - May) : Alubokharo and Rubio

Mid-Season (June - July) : Hale, Gaviota and Abundance

Late (July - August) : Shiro and Kelsey

DESCRIPTION OF CULTIVARS

President

It belongs to prune group of *Prunus domestica.* Fruits are large in size, skin is rich purple and flesh is deep yellow, juicy and sweet. It is self - unfruitful and hence requires a pollenizer.

Victoria

It is the most widely grown variety in European plum. Self - fruitful, frost resistant and heavy bearer. The fruit is purplish red with good quality.

Green Gage

It belongs to plum group called Reine Claude (hybrids of *Prunus domestica* x *Prunus insititia*). Fruits are medium in size, roundish in shape with greenish yellow skin. Flesh is firm and sweet.

Kelsey

It belonga to Japanese plum group (*Prunus salicina*). The fruits are heart shaped with greenish yellow skin and flesh. Flesh is firm but juicy with good quality.

Santa Rosa

It is also a Japanese plum. The fruit is large, purplish crimson. The flesh is amber with red near the skin. It is self fruitful and heavy bearing. It has got a characteristic flavour, a sour taste and thick skin. It is a popular variety grown throughout India.

Damson

It belongs to *Prunus insititia*. It is a late maturing heavy fruit bearer. Fruit is dark crimson with rich flavour. It is a self sterile variety.

Propagation

Plums do not come true to type from seed and hence its cultivars are propagated by budding on the seedlings or vegetatively propagated root-stocks obtained by means of layering or cutting. Under Indian condition, plums are shield budded on common peach seedlings of 1- 2 year old during November – January. Plums are propagated by 'T' budding in autumn or spring on root stocks of seedling origin or clonally propagated ones. Recently chip budding is found to be better. Scion – from one year old shoots from bearing trees. Myrobalan plum (*P. cerasiformae*) is commonly used rootstock in England and U.S.A. for European and Japanese plums.

Root Stocks Used

1. Myrobalan plum (*Prunus cerasifera):* It is the most widely used rootstock. Clones are Myrobalan 29 C, Myrobalan B 9 developed at East Mailing

Research Station, UK, presently known as Horticultural Research Institute, UK. Other clones are Myrobalan C, D and E.

2. Mariana plum (originated in Mariana, Texas, USA): Botanically a cross between *Prunus cerasifera* x *Prunus munsoniana.*
3. *Prunus domestica* selections: Hard wood cuttings and layers of Brompton and Brussel.
4. St. Julien stocks *(Prunus insitita)* selection.
5. American plum *(Prunus americana)*
6. Peach (*Prunus persica* propagated through seeds): Almond (*Prunus amygdalus* - seedlings).
8. Crosses of *P. amygdalus x P. persica.*

Root growing of stocks

A. Seeds

- Collected from fully ripe fruits.
- To break dormancy – stratified with moist sand for 45-50 days in 30-45 cm deep pits in cool and shady place.
- Sown in nursery buds – Dec. covered with mulch to protect from frost.
- One year old seedlings – used as R/S (root stock).

B. Clonal root stocks: Hardwood cuttings treated with 2500 ppm IBA and planted during January – February.

Preparation of Field

One to two year old budded plants are planted during spring before bud break under extreme winter condition. Under mild winter condition, autumn planting is recommended. Pits of size 60 cubic centimetres are dug at a spacing of 5 - 6 M either way and these pits are filled with top soil and compost (1:1). A spacing of 4.0 x 4.0 m is usually followed in South Indian conditions while 5 to 6.0 m spacing on either side is followed under North Indian conditions.

Planting

The plants should be set at depths in pits either equal to the depth to which the roots have gone in their original nursery or even slightly deeper than the original depth, and soil around the plant should be compacted. At the top, loose soil or FYM is spread to act as mulch. At the time of planting the plants should be headed back to 60 cm. large side branches below this point should be removed.

Trunk should be painted upto 15 cm from the headed top which will help to prevent sunburn and flat headed borer. Irrigation should be sufficient in the growing season to induce new growth and flower bud differentiation. The moisture stress may reduce the flower buds as well as fruit-set and the growing fruits are also affected. However, excess watering should be avoided as it will result in the loss of tree especially when peach is used as rootstock.

Manure and Fertilizers

An 8 - 10 year old plum plantation removes annually the following nutrients from the soil.

- N (Nitrogen) - 24 to 62 kg / ha
- P_2O_5 (Phosphorus pentoxide)- 4 to 13 kg / ha
- K_2O (Potassium oxide) - 23 to 62 kg / ha
- Ca (Calcium) - 4 to 12 kg / ha

Accordingly the periodical replacement of nutrients is essential. The leaf nutrient analysis is very useful to determine the fertilizer to be applied. An N : K ratio 1.4 : 1.0 should be maintained so as to get good yield. If excess nitrogen is applied to raise this rate to 2 : 1 then there will be yield reduction. Under Himachal Pradesh the following doses of fertilizers are recommended per tree. In South India, for bearing tree, annually during October – November, a dose of 30 kg of FYM and 1.5 kg of the fertilizer mixture (9:9: 9) are applied per plant.

Training and Pruning

The growing point is tipped at a height of 50-60cm to allow side shoots to develop. The trees are usually trained to modified central leader systems or the open centre system. Japanese plum cultivars are mostly adaptable to the open centre system.

Pruning is the prebearing period is usually kept light and corrective. In bearing plums, pruning is directed towards increase the size of fruits, preventing the breakage of branches as a result of heavy crops and promoting continuous new growth. Japanese plums require heavier annual pruning as compared to the European plums as the former one tends to overbear and may set such heavy crops that the fruit size is considerably reduced.

Cropping

The trees start flowering and fruiting 4-5 years after planting and a regular bearing is obtained after 10 years. The existing cultivars may be classified as self-sterile, partially self – fertile or self-fertile. Besides, flowers of plums exhibit hercogamy ie difference in the length of stamens and styles exist. Hence, plum cultivars grown commercially require pollination to set satisfactory crops. A cultivar is said to be self fruitful, if it exhibits an average of 30 per cent of fruitset which is considered adequate for a commercial crop. The cultivars that are self-unfruitful set only 1 to 2 percent. Hence, these cultivars should be interplanted with pollinisers. Most Japanese plum cultivars are self- unfruitful and require cross-pollination and some cultivars like 'Santa Rosa' and 'Beauty' are partially to fully self-fruitful and may bear heavier crops when planted among other cultivars. It is advisable to keep 5-10 beehives per hectare on the orchard periphery to increase the fruitset. Some plum cultivars tend to over bear and removal of part of the crop (thinning) enables the remaining crop to improve the size, quality and grade of plum fruits. Thinning is achieved by hand thinning and also by spraying growth regulators like NAA and 2,4,5-T at 25 and 50 ppm.

Intercultivation

Only the basins of the trees upto the extension of canopy are cleaned and the other inter space is grown either with a crop or sod (grass) so that they can serve as mulch. A semi non-conventional system in which 1 - 1.5 m strip with 15 cm high berms along the tree rows is kept clean by occasional cultivation or pre-emergence application of herbicide once a year. The inter row space can be covered with a permanent cover crop like self seeding clover or grass which has to be mowed periodically. Weed control near the tree basins can be done by using herbicides like simazine, dichlobenil, amitrole + diuron + MCPA without causing any injury to plum trees.

Growth Regulator Application

- Blossom thinning - 200 ppm ethephon and 100 ppm carbaryl.
- 500 ppm ethephon - a week before harvesting to improve fruit colour.
- 50 ppm triacontanol - after pit hardening stage – improves fruit size.

Fruitset

For satisfactory fruitset especially in self-unfruitful varieties pollenizers have to be inter planted. To increase the fruitset or to prevent excessive fruit drop application of GA_3 (Gibberrellins) 50 ppm + 2, 4 5 - T 10 ppm on the 10 th day after full bloom has to be done. Fruit thinning is another important operation in plum to get good sized fruits without sacrificing the total yield per tree. Spraying DNOC (6 - Di Nitro Ortho Cresol) 60 - 65 ml / 100 litres of water slightly before full bloom can effectively thin the fruits. Ethephon (Etherel) 200 ppm can be sprayed 2 weeks after full bloom which will help to thin the fruits as well as increase the individual fruit weight and thereby yield.

Harvesting

- Climacteric fruit.
- Picked at proper stage.
- Maturity indices
- Ex. Santa Rosa – days from full bloom to harvest.
- TSS – 16°brix.

Plums should be picked when mature enough to continue it's ripening process off the tree and plums grown for distant markets should be picked when firm ripe. Plums are picked by hand and packed in shallow crates. Each well grown tree can bear 25-30 kg of fruits. Fruits picked without pedicel and bruising and stem punctures should be avoided. Collected in picking baskets lined with gunny cloth and transported to grading and packing houses.

After full maturity, the fruits can be harvested before ripening and as far as possible the fruits should be harvested along with the fruit stalk (peduncle). Plums are packed in shallow crates (10 to 12. 5 cm deep) by placing to a maximum of 3 layers and should be cooled immediately to 0° C by which they can be kept without ripening at least for 12 days.

Yield - 60 - 70 kg fruits/tree

Postharvest Technology

Grading

3 grades based on fruit size

- Special - 42 mm and above
- Grade I - 36-42 mm
- Grade II - below 36 mm

Storage

- Can be stored for 2-4 weeks at 0°C with 85-90 % RH.
- Dipping in 4% calcium chloride for 2 minutes increases their storability without any reduction in firmness – for 12 days.

Nutritional Deficiency - Boron deficiency

- Misshaped fruits.
- Corky spots on the fruits along with cracking.
- Terminal buds fail to develop.
- Die back of shoots.
- Control – 0.1% boric acid during June.

PEACH

Botanical name: *Prunus persica*

Chromosome No.: 2n = 16

Family: Rosaceae

Origin: China

Introduction

Peach is a temperate juicy fruit of excellent quality. Its close relative which is actually a smooth skin mutant is called 'nectarine' (fruit skin is non - pubescent). Both of them are rich sources of sugar, protein, organic acids, vitamins and minerals. Used as table fruit, canning and for dehydration. The probable origin of peach is China where it would have been under cultivation even 3000 years back.

Uses

1. **Table purpose:** Yellow-fleshed, free from stone and regular bearer.
2. **Canning purpose:** Yellow flesh, free from stone, small pit, non-splitting, good symmetrical size and even maturity.
3. **Dehydration:** White fleshed, sweet peaches and free from stone.

Climatic and Soil Requirements

Peach is commercially grown in the world between 25° and 45° latitude above and below the equator. Just like any other temperate fruit it also requires a minimum amount of chilling (cumulative temperature hours below 7.2°C). However unlike other temperate fruits, the flower buds of peach are not hardy to resist the very low temperature, frost during spring or hailstorms and high humidity. After the completion of chilling requirement specific to varieties, the blooming time is dependent on post rest heat accumulation. At the time of fruit development, dry warm to hot summer climate is needed for better fruit development. Strong winds and hailstorms are unfavourable for peach cultivation. Mid elevation of hills (800 – 1500 M MSL) without soil freezing are better. Sandy loam soils rich in organic matter, with perfect drainage and at least 1 M depth are the best. The pH can be between 5.8 to 6.8 Very acidic or saline - alkaline soils are not fit for peach cultivation.

VARIETIES

Elberta

Fruits are large, round, pale yellow with red splash. Flesh is firm, yellow juicy, very sweet and free stone. It is a canning variety.

Hale

It is another commercial peach cultivar with large yellow fruits and free stone. It is good for both fresh market and canning.

Florda Sun

An early maturing cultivar developed in Florida. Fruits are round in shape with red blush on the surface. Fruit pulp is yellow, free stone, acidic sweet taste. It is a low chilling cultivars.

Florda Queen

Fruits are yellow with light red blush, round in shape. Flesh is firm, yellow, clingstone.

Red Globe

Fruits are medium in size attractive bright red in colour. Flesh is yellow firm melting with a small pit.

Alexander

Early season variety, Fruits are round with unequal sides, cavity deep, skin smooth, beetroot purple with some patches of pod green colour ; flesh soft with greenish white, colour, sweet, juicy aromatic and free stone.

Nectared 4

Mid season nectarine cultivar. Fruits are with good quality and colour.

Early Amber

It is a low chilling cultivar (350 hours). Fruits are with yellow skin and dark red blush. Flesh yellow, clingstone.

Sharbati

Late cultivar. Fruits round oblong medium in size, clingstone, pulp white soft and juicy. It is a heavy bearer.

Sunred

Low chilling (300 hrs) nectarine with small bright red fruits. Flesh yellow, firm and semifree stone.

Sunrise

A nectarine with globose fruits having medium thick, tough, mahagony red skin and yellow flesh.

Candor

Candor is a low chilling variety, can be cultivated in well drained sandy loam soil. It is the earliest variety suitable for home garden and large scale plantation. The fruits are available from mid May to first week of June. Since fruits are soft, they must be picked slightly before fully ripe stage and handled properly. They have yellow flesh with free stone, good flavour and aroma. It can be used fresh or canned or in dried form.

Florda prince

An early ripening vigorous variety, fruits are large with little or no tip, pubescent and semi - cling. External surface of fruits is red - blush with ground colour; flesh yellow firm with some red colour near to pit. Suitable for cultivation in sub tropical belt.

Earli Gande

A mid - season variety; fruits large with red blush having a medium pubescence; flesh colour yellow with a small amount of red colour near to pit. Suitable for subtropical region.

Propagation

Peaches are commercially propagated by budding (shield or ring method) and grafting (cleft method) on either seedling peach or plum or apricot rootstocks and on clonal interspecific hybrids of peach x apricot (*Prunus persica x Prunus amygdalus).* In the cultivar Shan-i-Punjab, the closer spacing of 6 x 3 x 3 m and the trees trained to 'Y' shaped structure produced higher fruit yield. The 'Y' shaped trees intercepted more quality fruits. Peaches are also multiplied clonally by hardwood cuttings, semi hard wood cuttings using treatment with growth regulators like IBA + NAA 5000 ppm by quick dip method.

Planting

The site is cleared off the bushes and weeds. In hill slopes necessary precaution should be taken to facilitate soil conservation and moisture conservation by forming contour bunding at the same time taking utmost care for perfect drainage. Since peach is the most sensitive temperate fruit tree to water logging. Pits of 1 cubic metre size can be dug at a spacing of 4. 5 M x 1. 5 M during September - October and filled with 30 – 40 kg of well rotten FYM. At the time of planting 40 - 50 g of aldrin or BHC 10% should be added to each pit. Normally peach seedlings will not establish in areas where origin benzaldehyde through decomposition of amygdalin a chemical present in cells of peach root bark. If the old roots of previous peach trees are completely decomposed such problem will not occur. Normally 1 year old plants are planted at a spacing of 4.5 M x 1.5 M. In hedge row planting the plants are planted at a spacing of 2 M x 1 M and alternate rows should be removed by fifth year so that the final spacing will be 4 M x 1 M such high density planting will help to fetch more yield in the initial years. Planting of peach is commonly done in late winter or early spring where there is freezing winter. In mild winter climate, planting can be done during late full or early winter, so that the roots get sufficient time to establish before the top starts to grow. In the centre of the pit, soil should be removed to same depth as the plant was in nursery and after planting the soil should be pressed around. Then the plants are irrigated and then the bases are mulched with hay or oak leaves or FYM to a thickness of 10 - 15 cm.

Manures and Manuring

A peach crop producing 25 t / ha removes 140 kg N, 10 kg P and 80 kg K, 75 kg Ca, and 18 kg Mg. Normally for a bearing peach tree 400 g N, 100 g P and 400 kg K is recommended per pear. Foliar application of 0.5 % Zn sulphate is recommended to increase the yield. Normally ¾ th of the N and full dose of P and K should be applied in trenches of 20 – 25 cm width, 10 - 15 cm depth made beneath the trees at a distance of 1 - 2 M away from the trunk just prior to dormancy. Remaining ¼ th of Nitrogen should be given immediately after fruit harvest. On the basis of tissue analysis, it has been estimated that for highyield the tree nutrient levels should be 3.37 % N, 0. 2 % P, 1. 46 % K, 1.16 % Ca and 0.35 % Mg.

Inter Cultivation

Cow pea can be the best inter crop in peach orchards during early growth periods. Soybean and Pine apple also serve as good inter crops. But permanent sod covers should be avoided as they result in poor tree growth. Paraquat @ 570 g / ha applied thrice at monthly interval can effectively check the weeds. Application of Atrazine 5 kg and Oxyfluorfen 0.2 kg / ha can effectively control both dicot and monocot weeds. Peach tree will come to bearing in second or third year after planting. Most of the varieties are self fruitful except J. H. Hale, Elberta, Candoka and Chinese cling. For self sterile varieties a pollenising variety should be inter planted viz., Two rows of self sterile varieties should be alternated with two pollenising variety. Fruit thinning can be done by reducing the crop load through adequate pruning in the dormant season itself. It can also achieved by spraying ethephon 5 ppm when the fruits are 20 - 25 mm in size.

Training and Pruning

Generally one year old whip without a single branch is plant in the dormant season. Immediately after planting the tip is headed back to 60 - 75 cm above the ground level. In the following summer most of the buds on main trunk will sprout. Buds sprouting upto 30 cm from the ground level should be nipped off soon after their appearance, to develop a smooth clean trunk. Above that 3 or 4 well spaced buds projecting in opposite direction should be allowed. In the second summer the unwanted buds should be again pinched off. In winter, all the weak, diseased, dead and interfering shoots should be removed. Laterals which grow towards the centre of the tree or vertically should be completely thinned out. The selected branches are then headed back to a bud projecting in the outer direction. This system is called modified leader system and will be useful for maximum interception of abundant sunlight. Peach trees bear fruit bud laterally on previous season's growth.

For bearing trees regular pruning which has to be done to maintain a balance between vegetative and reproductive parts of the tree. During the early bearing, the selected shoots should be lightly pruned by removing 20 - 30 percent of linear growth. After 7 - 8 years of bearings the natural vigour of the tree is diminished, then corrective pruning should be resorted to. This is done by removal of linear extension growth to the extent of 70 - 80 percent in 50 % of the shoots. In the remaining 50 % of shoots light pruning is done to induce early bearing. This 50 % shoots can be severely or moderately pruned during next season. Thus moderate and light pruning should be combined in the alternative branches of the single tree.

Harvest and Yield

Peach fruits should be harvested after full maturity if harvested hard and green it will not attain proper eating quality even after artificial ripening. At 'firm' stage there is a slight colour change and fruit yields slightly to pressure. This stage is suitable for a distant market however it is slightly inferior quality on ripening. At 'firm ripe' stage the fruit colour changes and fruit can be pressed in a cupped hand but still it has not attained the best eating quality. At this stage it can be stored for a longer period. At 'tree - ripe' stage the fruit is fully ripe and yields readily to pressure. At this stage the fruit has attained the best eating quality. But the fruits can be sold only in local market immediately. They cannot be either stored or transported to distant market. 'Soft - ripe' is a stage when the fruit has lost its firmness. It has high dessert quality, but should be consumed immediately. The average yield ranges from 7 - 10 tonnes / ha / year. Under high density planting and Tatura trellis planting, yields as high as 70 - 80 tonnes / ha have been recorded even at 3rd to 4th year of planting.

POSTHARVEST TECHNOLOGY

Grades of Peach

3 grades based on fruit size (mm).

- Special - 55-66 mm.
- Grade 1 - 46-55 mm.
- Grade 2 - < 46 mm.

Storage

Peaches and nectarines can be stored for even 28 - 36 days at a temperature of 0°C and 85 - 90% RH.

❑❑❑

CHAPTER

21

IMPORTANCE AND AREA OF PLANTATION CROPS

Importance of Plantation Crops

The term Plantation crops refers to those crops which are cultivated on an extensive scale in a large contiguous area, owned and managed by an Individual or a company. The crops include tea, coffee, rubber, cocoa, coconut, arecanut, oil palm, palmyrah, cashew, cinchona etc. These plantation crops are high value commercial crops of greater economic importance and play a vital role in our Indian economy. The main draw back with this sector: of crops in India is that major portion of the area is of small holdings (except Tea) which hinders the adoption of intensive cultivation. In the case of coffee 97.13 per cent of the growers have holdings below ten hectares and in Rubber, 82 per cent of the total area is of small holdings having an average size of 0.5 ha.

The Economic Importance of these Crops are:

1. They contribute to national economy by way of export earnings. These crops occupy less than 2 per cent of the total cultivated area (i.e. 3.82 per cent of total crop land) but they generate an income of around Rs. 16,000 million or about 12.72 per cent of the total export earnings of ail commodities or 75 per cent of total earnings from the export of agricultural produces.
2. India is the leading country in the total production of certain plantation crops in the world. For instance, our production meets the share of 47 per cent in tea and 66 per cent in each of cashew and arecanut,
3. Plantation industry provides direct as well as indirect employment lo many millions of people. For instance, tea industry offers direct employment to 10 lakhs and indirect employment to 10 lakh people, while-cashew processing factories alone provide employment to 3 lakhs people besides 2 lakhs farmers are employed in cashew cultivation.

4. Plantation industry supports many by-product industries and also many rural industries. For example, coconut husk is used to produce coir fiber annually to a tune of 2,19,600 tones in India.
5. These crops help to conserve the soil and ecosystem. Tea planted in hill slopes and cashew in barrel and waste lands protect the land from soil erosion during the rainy season or due to heavy winds.

Area of Plantation Crops in India

The area and production is more in Kerala (1073000.7 ha and 4359000.9 MT) followed by Karnataka (679000.2 ha and 1401000.0 MT), Tamil Nadu (503000.0 ha and 3810.6 MT), Andhra Pradesh (289000.0 ha/and 1021000.2 MT).

❑❑❑

CHAPTER

22

INSTITUTIONS INVOLVED IN PLANTATION CROPS

Central Plantation Crops Research Institute (CPCRI) - Kasaragod, Kerala

Central Plantation Crops Research Institute (CPCRI) was established in 1970 as one of the agricultural research institutes in the National Agricultural Research System (NARS) under the Indian Council of Agricultural Research (ICAR). The present organizational set up of the Institute with headquarter at Kasaragod, Kerala, Regional Stations at Kayamkulam (Kerala), Vittal (Karnataka) and Minicoy (Lakshadweep islands) and Research Centres at Kahikuchi (Assam), Mohitnagar (West Bengal) and Kidu (Karnataka) cater to the research needs of the mandate crops at the National level. The research programmes of the institute is organized under five divisions viz. Crop Improvement, Crop Production, Crop Protection, Physiology, Biochemistry and Post-Harvest Technology and Social Sciences. The Institute also holds the International Coconut Gene Bank for South Asia at Kidu in Karnataka. The Institute also serves as the headquarter of All India Co-ordinated Research Project (AICRP) on Palms with 13 centres on coconut, 7 centres on oil palm and 2 centres on palmyrah.

Directorate of Oil Palm Research - Pedavegi, Andhra Pradesh

Research on oil palm under irrigated conditions is practically a new start and the Indian Council of Agricultural Research established the National Research Centre for Oil Palm at Pedavegi in West Godavari district of Andhra Pradesh on February 19, 1995. In February, 2009 the institute was upgraded as Directorate of Oil Palm Research (DOPR) to cater to the location specific programmes across the country. DOPR serves as a centre for conducting and coordinating research on all aspects of oil palm conservation, improvement, production, protection, post-harvest technology and transfer of technology. This institute has a research centre at Palode near Thiruvananthapuram, Kerala.

Directorate of Cashew Research - Puttur, Karnataka

Research on cashew was first initiated in the early 1950s. Indian Council of Agricultural Research (ICAR), sanctioned ad hoc schemes for Research Centres located at Kottarakkara (Kerala), Ullal (Karnataka), Bapatla (Andhra Pradesh), Daregaon (Assam) and Vengurla (Maharashtra). In 1971, ICAR also sanctioned All India Coordinated Spices and Cashew Improvement Project (AICS and CIP) with its Head Quarters located at CPCRI, Kasaragod. The CPCRI Regional Station, Vittal (Karnataka) was given the mandate to carry out research work on cashew while four University Centres (Baptala, Vridhachalam, Anakkayam and Vengurla) were assigned the research component on cashew under AICS and CIP. During the V and VI plan three more centres (Bhubaneswar, Jhargram and Chintamani) came under the fold of AICS and CIP and with shifting of work of Anakkayam centre to Madakkathara. The recommendations made by the Quinquennial Review Team (QRT) constituted by ICAR in 1982, working group on Agricultural Research and Education constituted by the Planning Commission for VII Plan Proposals and the Task Force on Horticulture constituted by ICAR had resulted in the establishment of National Research Centre for Cashew at Puttur on 18th June 1986 which was upgraded and renamed by ICAR in 2009 under XI Plan as Directorate of Cashew Research (DCR). Subsequent to the bifurcation of AICS and CIP, the headquarters of All India Coordinated Research Project on Cashew was shifted to DCR, Puttur. At present, this Coordinated Research Project is operating in ten centres and a sub centre distributed in major cashew growing areas of the country. There are also three co-operating centres

Indian Institute of Horticultural Research - Bangalore, Karnataka

The Institute spread its sphere of Research activities to the length and breadth of the Nation by establishing its experimental stations at Lucknow, Nagpur, Ranchi, Godhra, Chettalli and Gonikopal. Over the years these experiment stations have grown in size and today they stand as independent institutes, however, retaining the Chettalli and Gonicoppal under its fold. As of now, the IIHR has its main research station at Hessaraghatta, Bangalore with 263 ha of land and Regional experiment stations at Bhuvaneshwar in Orissa and Chettalli in Karnataka with two Krishi Vigyan Kendras both located in Karnataka state at Gonikopal in Kodagu and Hirehalli in Tumkur districts. Apart from this the Project Coordinating Cell of the All India Coordinated Research Project on Tropical Fruits is also located at the institute at Bangalore.

Coconut Development Board - Kochi, Kerala

Coconut Development Board is a statutory body established by the Government of India for the integrated development of coconut production and utilization in the country with focus on productivity increase and product diversification. The Board which came into existence on 12th January 1981, functions under the administrative control of the Ministry of Agriculture, Government of India, with its Headquarters at Kochi in Kerala and Regional Offices at Bangalore in Karnataka, Chennai in Tamil Nadu and Guwahati in Assam. There are six State Centres situated at Bhubaneswar in Orissa, Calcutta in West Bengal, Patna in Bihar, Thane in Maharashtra, Hyderabad in Andhra Pradesh and Port Blair in the Union Territory of Andaman & Nicobar Islands. The Board has 9 Demonstration cum Seed Production (DSP) Farms in different locations of the country and now 7 farms are maintained. A Market Development cum Information Centre has established in Delhi. The Board has set up a Technology Development Centre at Vazhakulam near Aluva in Kerala.

Rubber Research Institute of India (RRII) - Kottayam, Kerala

The Rubber Board is a statutory body constituted by the Government of India, under the Rubber Act 1947, for the overall development of the rubber industry in the country. Commercial cultivation of natural rubber was introduced in India by the British, although the experimental efforts to grow rubber on a commercial scale in India were initiated as early as 1873 at the Botanical Gardens, Calcutta. The first commercial Hevea plantations in India were established at Thattekadu in 1902. The importance of rubber production in India from strategic and security reasons had been realized by the government during the Second World War period. The rubber growers in India were encouraged to produce the maximum rubber required for the use during war. After the war, there were growing demands from the growers for setting up a permanent organization to look after the interests of the industry. Thereupon the government set up an ad-hoc committee in 1945 to study the situation and to make appropriate recommendation. On the recommendation of this ad-hoc committee, the government passed the Rubber (Production and Marketing) Act, 1947, on 18th April 1947, and the "Indian Rubber Board" was constituted forthwith. The Rubber Production and Marketing (Amendment) Act, 1954, amended the name of the Board as "The Rubber Board".

Directorate of cashewnut and cocoa development - Kochi, Kerala

The Development of Cashew was first introduced in the early 50's and was hand in glove with the Research and Development of the ICAR looked after by Indian

Cashew nut and Spices Committee till the beginning of 3rd plan period. Cashew Development was taking place from the beginning through adhoc schemes sanctioned by the erstwhile Cashew nut and Spices Committee. In the post independent era from the beginning of 3rd Plan, no central agency was there to devote an individual attention for the development of cashew in India. The Directorate of Cashew nut Development established in 1966 as a primary field functionary under the Union Ministry of Agriculture gave a greater impetus for the development of cashew in a more scientifically oriented manner. This marked the first step towards the integration and co-ordination of Cashew Development in association with developmental agencies of States and Research Institutes. Now cashew development and research go hand in hand conceiving whatever technological advancements taken place in the research front becoming an integral part of the developmental efforts. The Directorate of Cashew nut Development which was handling only cashew got the mandate for development of cocoa in 1997. Though cashew and cocoa enjoys different parameters, both are economical cash crops of the country and in 1997-98 the Directorate of Cashew nut and Cocoa Development started driving both the crops on similar tracks

Coffee Board - Karnataka

The saga of Indian coffee began on a humble note, with planting of 'Seven seeds' of 'Mocha' during 1600 AD by the legendary holy saint Baba Budan, in the courtyard of his hermitage on 'Baba Budan Giris' in Karnataka. For quite a considerable period, the plants remained as a garden curiosity and spread slowly as back yard plantings. It was during 18th century that the commercial plantations of coffee were started, thanks to the success of British entrepreneurs in conquering the hostile forest terrain in south India. Since then, Indian coffee industry has made rapid strides and earned a distinct identity in the coffee map of the world. Century that the commercial plantations of coffee were started, thanks to the success of British entrepreneurs in conquering the hostile forest terrain in south India. Since then, Indian coffee industry has made rapid strides and earned a distinct identity in the coffee map of the world.

Tea Board - Kolkata, West Bengal

Tea is one of the industries, which by an Act of Parliament comes under the control of the Union Govt. The genesis of the Tea Board India dates back to 1903 when the Indian Tea Cess Bill was passed. The Bill provided for levying a cess on tea exports - the proceeds of which were to be used for the promotion of Indian tea both within and outside India. The present Tea Board set up under section 4 of the Tea Act 1953 was constituted on 1st April 1954. It has succeeded the Central Tea Board and the Indian Tea Licensing Committee which functioned respectively under the Central Tea Board Act, 1949 and the Indian Tea Control Act, 1938 which were repealed. The activities of the two previous bodies had been confined largely to regulation of tea cultivation and export of tea as required by the International Tea Agreement then in force, and promotion of tea Consumption.

❑❑❑

CHAPTER

23

COCONUT

Botanical name: *Cocos nucifera* L.

Chromosome No.: 2n = 32

Family: Palmae

Provides food, fuel and timber. Every part of the tree is being utilized for some purpose or other and hence, it is called Kalpavriksha meaning tree of heaven which provides all the necessities of life.

Origin - South East Asia

Botany

The coconut belongs to the family of palms (Palmae), included under the lower group of flowering plants known as the *monocotyledons*. It occupies a conspicuous position in the vegetative kingdom owing to certain characteristic features of the palms, viz., comparatively slow growth, and unusual thickening at the base of the trunk giving mechanical rigidity, generally unbranched, erect, cylindrical, pillar like stem, covered with heavy scars of the old leaf bases, a compact magnificent crown of gigantic, feathery, glossy thick-cuticled leaves, sheathed at base providing firm attachment to the stem oriented at the top of the trunk rendering in a sheath collectively known as the spadix; the absence of tap root and the continually growing terminal bud commonly known as the *'Cabbage'*.

The genus *Cocos* is a monotypic one containing the only species, *Cocos nucifera* Linn. It is a tall, stately unbranched palm growing to a height of 12 to 24 m. The trunk is stout, flexuosus, annulate, rarely stands vertically, but makes a gradual curve, rising from a swollen base surrounded by a mass of adventitious roots. The stem is marked by rings of leaf scars, which are often not prominent at the base.

The coconut palm has an adventitious root system of monocots producing numerous thick roots from the base of the stem almost throughout its life. The roots are localized generally at the lower most region of the stem, which has been termed the 'bole'.

Leaves large, long pinnatisect, borne on the crown; leaf-lets equidistant, 60 to 90 cm long, narrow, tapering, linear, lanceolate, coriaceous, flaccid; petioles stout, 90 to 150 cm long.

Spadix is 1.2 to 1.8 m long, stout, etect, straw or orange coloured, androgynous, simply branched, branches (spikes) bear one or more female flowers often between two male flowers towards their bases, and several males above; spathes 2, outer short, inner 60 to 90 cm long, oblong hard, splitting lengthwise. The palm is monoecious with female flowers relatively a few, male flowers are numerous, small, unsymmetric, sweet scented; sepals small, valvate, petals about 6.4 mm long, oblong, acute, valvate; stamens 6, filaments sabulate, anthers about 2.5 cm long, globose, bracteolate, perianth greatly acrescent, round, concave, imbricate petals are shorter than the sepals, convolute with imbricate tips; disc annular. Ovary is tricarpic, usually one ovuled sub-basilar.

Fruit is large 20 to 30 cm long, trigonolly obovoid or subglobose, three sided, and drupe one seeded. The outer layers of the pericarp are thick and fibrous. The inner layer (endocarp or shell) is very hard, horny or stone with three basal pores or marks representing the remains of 3 carpels or the loci of the ovary, two of which have become obliterated. Under one of these, lies the embryo. The thin testa cohering to the endocarp is lined with white albuminous endosperm (meat) enclosing a large cavity, partially filled with sweet fluid.

Climate and Soil

The coconut palm is found to grow under varying climatic and soil conditions. It is essentially a tropical plant. The palms tolerate wide range in intensity and distribution of rainfall. However, a rainfall of about 200 cm per year and well distributed throughout the year is the best for proper growth and maximum yield.

Coconut is adaptable to a wide range of soil conditions, from light sandy soils to heaviest clays with a pH ranging from 5.2 to 8.0.

Cultivars and Hybrids

Coconut palms are broadly classified into two groups, the tall and dwarf. The tall cultivars are the common types that occur throughout the world.

Variety	Main Characteristics
Talls	Thick stem with swollen base (bole). Late flowering (5–6 year from out planting). Little or no overlapping of male and female phases of an inflorescence encouraging out crossing
Dwarfs	Slender stem with short internodes. Bole slight or absent. Early flowering (3 year from out planting). Considerable overlapping of male and female phases of an inflorescence resulting in self-pollination

Tall Varieties

- West Coast Tall
- East Coast Tall
- Chandrakalpa or Lakshadweep Ordinary (LCT)
- Phillippines Ordinary Kerachandra)
- VPM - 3 (Andaman Ordinary)
  - Parentage – Selection from Andaman Ordinry
  - Year of release – 1994
  - Annual nut yield / palm – 92
  - Tall variety
  - 63 months for first flowering
  - Big oblong nuts
  - Oil content 70%
  - Copra content 176g/nut
  - Drought tolerant
- Aliyar Nagar 1 (ALR 1)
  - Time take for bearing: 5 years
  - Average Yield: 126 nuts / palm / year
  - Copra content: 131 gram / nut
  - Oil content: 66.5 per cent
  - Drought tolerant suitable for Rainfed and Irrigated condition
  - Selection from Arasampatti Tall (Dharmapuri Dist)
  - One year earlier flowering compared to WCT, ECT and VPM 3
  - Moderately tolerant to major coconut pests.
  - Released by Coconut Research Station, Aliyar Nagar, Tamil Nadu during 2002

- Aliyar Nagar 2 (ALR 2)
  - Parentage - Selection from Tiptur Tall
  - Yield - Average Yield: 109 nuts/palm /year
  - Annual yield: 18988 nuts/ha/year (12 % over ECT and 99% over VPM 3)
  - Nut bearing in 5½ years
  - Regular bearing habit
  - 12 inflorescences per year
  - Weight of copra - 135 g/nut
  - 2.57 tonnes of copra per hectare
  - 7400 nuts required to make one tonne of copra
  - Oil content 64.7 percent
  - Possesses drought tolerance
  - Moderately resistant to rhinoceros beetle, red palm weevil and leaf blight
- Tiptur Tall
- Kera Sagara (Seychelles)
- Phillipines Tall (Chandrathara)
- Assam Tall (Kamaroopa)
- Kalpadhenu
- Kalpa Pratiba
- Kalpa Mitra
- Kerakeralam

Dwarf Varieties (Tender coconut)

- Chowghat Orange Dwarf (COD)
- Chowghat Green Dwarf (CGD)
- Chowghat Yellow Dwarf (CYD)
- Malayan Yellow Dwarf
- Malayan Orange Dwarf
- Gangabandom
- Strait Settlement Dwarf Green
- Kalpa Raksha

Hybrids

- Kerasankara (WCT x COD)
- Chandrasankara (COD x WCT)
- Chandralaksha (LCT x COD)
- Keraganga (WCT x GBGD)
- Lakshaganga (LCT x GBGD)
- Anandaganga (ADOT x GBGD)
- Kerasree (WCT x MYD)
- Kerasoubhagya (WCT x SSAT)
- VHC 1 (ECT x MGD)
  - Parentage – East Coast Tall X Malayan Green Dwarf
  - Year of release – 1982
  - Annual nut yield / palm – 115
  - T X D hybrid
  - 40 months for first flowering
  - Oblong nuts
  - Oil content 68.6%
  - Copra content 142g/nut
- VHC 2 (ECT x MYD)
  - Parentage – East Coast Tall X Malayan YellowDwarf
  - Year of release – 1988
  - Annual nut yield / palm – 142
  - T X D hybrid
  - 43 months for first flowering
  - Medium to big oblong nuts
  - Oil content 70.2%
  - Copra content 146g/nut
- VHC 3 (ECT x MOD)
  - Parentage – East Coast Tall X Malayan Orange Dwarf
  - Year of release – 2000
  - Annual nut yield / palm – 156
  - T X D hybrid
  - 46 months for first flowering

- Medium to big oblong nuts
- Oil content 70.2%
- Copra content 162 g/nut

- Gadavani Ganda (East Coast Tall X Gangabandom)
- Kalpa Samrudhi
- Kalpha Shankara (Chowghat Green Dwarf X West Coast Tall)

Selection of Mother Palm

In the selection of mother palms, the following characters are looked for:

1. The palm should be a regular bearer and should give an annual yield of about 60 nuts under rainfed condition. The copra content per nut should be around 150g.
2. The crown should be spherical or semi-spherical drooping or erect crown should be avoided, and should have at least 30 freely opened leaves.
3. The length of petiole and bunch stalk should be short and stout.
4. The nuts should be medium in size and nearly round or spherical or oblong.
5. The bunches should have a preponderance of heavy nuts.
6. The trees producing barren nuts should be discarded.
7. The palms should be between the age group of 25-60 years.
8. The palms growing close to house, cattle sheds, compost heaps water holes etc. may be avoided as it is difficult to identify inherently good trees.

Preparation of Land and Planting

Normally a pit size of 1.0 x 1.0 x 1.0 m is dug and filled upto 50 cm depth with sand and powdered cow dung.

Spacing

A spacing 7.5 to 9.0 m may be adopted depending on the crown size. This will accommodate 177 to 124 palms per ha under the square system of planting. In well drained soils, seedlings can be transplanted with the beginning of south-west monsoon. If irrigation facilities are available, it is advisable to take up planting at least a month before the monsoon sets in so that the seedlings get well established before the onset of heavy rains.

Care of Young Palms

The transplanted seedlings should be shaded and irrigated properly especially during the summer months. The pits should be cleared of weeds periodically. Soil washed down by the rains and covering the collar of the seedlings should also be removed. The pits should be gradually filled up as the seedlings grow.

Manuring

Age (Years)	FYM (kg/tree)	Urea (kg/tree)	Super Phosphate (kg/tree)	Muriate of Potash (kg/tree)
1	10	0.308 (140 g N)	0.500 (80 g P_2O_5)	0.480 (300 g K_2O)
2	20	0.616 (280 g N)	1.000 (160 g P_2O_5)	0.960 (600 g K_2O)
3	30	0.924 (420 g N)	1.500 (240 g P_2O_5)	1.440 (900 g K_2O)
4	40	1.23 (560 g N)	2.000 (320 g P_2O_5)	1.920 (1200 g K_2O)
5th year onwards	50	1.23 (560 g N)	2.000 (320 g P_2O_5)	1.920 (1200 g K_2O)

Apply manures and fertilizers in circular basins of 1.8 m from the base of the palm, incorporate and irrigate. The fertilizers may applied in two split doses, in June – July and in December to January.

TNAU Coconut Tonic

For nut bearing coconut, root feed TNAU coconut tonic @ 200ml/palm once in six months.

Bio-fertilizer Recommendation

Mix 50 g of *Azospirillum*, 50 g of Phosphobacteria (or) 100 g *Azophos* and 50 g of VAM in sufficient quantity of compost of FYM and apply near feeding roots once in 6 months / palm starting from planting. Don't mix with chemical fertilizers and pesticides

Organic Recycling

Any one of the green manure crops like sunnhemp, Calapagonium or Daincha may be sown and ploughed *in situ* at the time of flowering as a substitute of compost to be applied. Sow sunnhemp @ 50 g/palm in the basin and incorporate before flowering. Coir pith compost/vermicompost made from coir pith /coconut leaves / other wastes from coconut grove can be applied.

Irrigation

The coconut palm responds to summer irrigation. Production of female flowers and setting percentage increases considerably due to irrigation, 2 cm irrigation once in 4-5 days during December-May is beneficial in sandy loam soils. Coconut husks or dusts can be added in pits to conserve moisture.

Inter Cultivation

Tillage operations like digging the garden with mammutty (spade), ploughing are beneficial to the trees. Method of intercultivation will depend upon local conditions, availability of labour, size of holding, soil type, topography and distribution of rainfall. Cover cropping is recommended to prevent soil erosion in coconut gardens. Leguminous crops such as *Mimosa invisa, Stylosanthes gracilis* and *Calapogonium mucunoides* are generally recommended. Green manure crops like sunhemp (*Crotolaria juncea*) and *kolinji* (*Tephrosia purpurea)* are also raised and ploughed in during August-September.

Inter and Mixed Cropping

A variety of intercrops like pineapple, banana, elephant foot yam, groundnut, chillies, sweet potato and tapioca can be raised in coconut gardens after the palms attain a height of 5 to 6 metres. In older plantations, cocoa, pepper, cinnamon, clove and nutmeg can be grown as mixed crops.

Harvesting

Harvesting is done at 45 days interval during summer months and at 60 days interval during the rainy season. Hence, 35 percent of the total nut is obtained during the hot months i.e., from March-May and least crop is obtained during rainy months. Harvesting is done by climbing the tree.

Yield: 60-80 nuts per palm per year

SPECIAL PROBLEM

Shedding of Buttons and Immature Nuts and Production of Barren Nuts

These are some of the important problems of coconut cultivation, which adversely affect the yield to a large extent. The probable causes which may be attributed are the attack of pests and diseases, nutritional deficiencies, unfavourable soil and climate conditions, defects in pollination and fertilization and the limited capacity of the tree to bear fruits.

Production of Barren Nuts

The phenomenon of the occurrence of barren nuts (without or with imperfectly developed kernel) is as ancient as the cultivation of coconut. From the detailed investigations carried out on this problem at Kasaragod, Kerala, it was revealed that in the coconut plantations certain trees produce a large number of barren nuts. The nuts are generally oblong in shape and the quality of husk produced is very much less as compared to the normal ones. The embryo in the barren nuts is mostly absent or when present, it is in varying stages of decay. Often, these nuts are seen with the shell and kernel improperly developed. Fungal infection is also sometimes noticed in the embryo, resulting in the decay of the kernel and loss of water inside. Of the different types of barren nuts met with, those with cracking of shell are relatively more common than the other types.

Preliminary trials carried out at the Central Plantation Crops Institute Research, Kasaragod, revealed that barrenness could be induced by keeping female flowers unfertilized indicating thereby that defective fertilization may be one of the possible causes of barren nut production. Preliminary work done at Kasaragod to determine the possibility of reducing the incidence of barren nuts through heavy manuring with 2.722 kg ammonium sulphate, 1.814 kg of bone meal, 27.215 kg ash and 90.718 kg of green mature per tree failed to show any distinct beneficial effect. Genetic causes are also attributed to button shedding. Although investigation made on the phenomenon failed to establish conclusive reasons for the occurrence of barren nuts. It is not however, improbable that excessive bearing may be one of the causes.

Plant Protection

The major pests and diseases affecting the coconut palm are furnished below:-

Pests / diseases	Control measures
Rhinoceros beetle (*Orycetes rhinoceros)*	Proper disposal of decaying organic debris extracting the beetle with a hook mechanically, filling the inner most three or four leaf axils of palm with a mixture of 5% BHC dust and sand (1:1), three times a year.
Red Palm weevil *(Rhynchophorus ferrugineus)*	Injection of Pyrethrin piperonyl butomide (Pyrocon-E) 10ml in one litre of water per palm into the trunk through a hole above the infested portion.
Diseases	
Bud rot *(Phytophthora palmivora)*	Application of Bordeaux paste in the cut portion of the infected tissues at early stage of infection itself.
Root (wilt) diseases	
Thanjavur wilt *(Gonoderma lucidum)*	Apply 5 kg of neem cake per year, addition of organic matter, providing irrigation could check the spread of disease.

Value Added Product

Coconut is the most beautiful and useful of palms known to mankind. Coconut is unique, producing a host of products. Presently India is having an area of 18, 95,000 ha with production of 15,730 million nuts.

Products of coconut can be broadly classified into 3 categories. They are: food products, commercial products, coconut shell and miscellaneous products. Farm level processing either individually or collectively can result in higher income. When coconut is converted to milling copra there is value addition of about 25% and into edible ball copra it is 35-40%. Value addition is nearly 22% by marketing tender nut in place of mature nut.

Coconut Milk Powder

Two types of health mixes can be prepared from coconut milk powder viz., therapeutic health drink and malted health drink. For therapeutic health drink, initial content of chemical constituents like moisture, protein, energy, fat, fiber, calcium, phosphorus, iron, free fatty acid and peroxide value were 7.68%, 9.12 g, 335 Kcal, 1.8 g, 1.39%, 430 mg, 222 mg, 1.76 mg, 0.23%, 0.39 meq respectively / 100 g of product and for malted health drink were 9.09%, 15.66 g, 323 Kcal, 2.9 g, 1.62%, 566 mg, 219 mg, 1.95 mg, 0.32%, 0.54 meq respectively / 100 g of product. Cost of 100 g of health mixes is Rs. 25/-.

Coconut Syrup

Coconut syrup is prepared by adding an equal quantity of sugar and 0.05% citric acid to coconut milk, followed by steam-cooking to a total solids content of 65-68%. It is used as a drink and a bread-spread.

Coconut Honey

Coconut honey is prepared in similar way from coconut milk by adding brown sugar and glucose.

Desiccated Coconut

It is the dried, disintegrated coconut meat of fully matured nut stored for a month before de-husking. This makes de-shelling easier and whole kernel is removed from shell. Brown testa covering the kernel is removed and then kernel, washed, sterilized, disintegrated and desiccated at 77-82°C to a final moisture content of 2-2.5%, followed by cooling.

Coconut Flour

Coconut grating from which fat is removed partially by hydraulic pressing or fibrous residues after extracting milk are dried and powdered to obtain coconut flour. Coconut flour contains 31% carbohydrate and 43% fibre. It is utilized in bakery and confectionery products.

Dietary Fibre

Spent coconut residue is a good source of dietary fibre especially insoluble fibre. High fibre content especially insoluble fibre offers scope for its utilization as a dietary component against constipation, obesity and for diabetic patients.

Sensory quality of products prepared out of dietary fibre *viz.* biscuits, coconut burfi, coconut chutney (dry), sambar and traditional Indian sweets) was good and overall acceptability was very high. Substitution of rice by 10% with coconut flour in dosa and wheat by 20% in chappatis has got acceptability.

Toddy Products

Toddy is a sweet juice containing sucrose as the main constituent obtained by tapping unopened spadix. Dwarf palms are highly suitable for tapping due to their short stature and toddy is sweeter but yield is less than tall palms. Tapping toddy increases yield of poor-bearers.

Fresh toddy, if not collected under sterilized conditions, is fermented rapidly. Sugar turns into alcohol (5.8%) and product collected by distillation is strong liquor known as Arrack. Other products obtained are jaggery and refined sugar by evaporation and coconut vinegar by fermentation of toddy.

Virgin Coconut Oil

Virgin coconut oil is naturally processed product of fresh coconut meat (kernel) either directly or through extraction of coconut milk. It is colourless (water white) and has a mild to intense coconut scent and is rich in lauric acid (47-53 %).

It also contains vitamin E (tocopherols) at 5 mg / kg. Virgin coconut oil in its pure form can be directly consumed and has a mild, sweet scent. It is an excellent and exotic salad dressing, a good food seasoning and serves as healthy cooking oil because it is highly resistant to heat. This also has wider uses in functional foods, pharmaceuticals and cosmetics.

Coir and Coconut Fibre

It is an important product obtained from fruit husk. Husk forms 35-45% of whole nut when ripe of which 30% constitutes coir fibre and 70% are pith and outer skin. Lignin and cellulose are the major constituents of pure fibre. There are 2 types of fibres, white fibre and brown fibre.

White Fibre

It is extracted from green husks after retting in natural water for 6-10 months until fibre becomes loose and soft. When facilities for natural wetting are lacking, it can be subjected to mechanical decortication and extraction. Nuts of 11 months old are best suited for extraction of white fibre. White fibre is long and fine, suitable for spinning into coir yarn for making ropes and mats. Production is mainly confined to Kerala.

Brown Fibre

Brown fibre is extracted from ripe dry husks by mechanical de-fibring process. Two types of brown fibre are bristly fibres and mattress fibres. Bristly fibre is short and used for making brush, whereas mattress fibre is long and used for stuffing purpose in upholstery and mattress.

Coir Pith or Dust

A waste product of coir industry known as cocopeat, it is used as a mulch, for making briquettes with high calorific value, as a manure after composting with 'pith plus' a formulation containing *Pleurotus sajor caju*. It can also be used for production of biogas, lightweight building bricks and as a soil conditioner.

Coconut Shell-based Products

Coconut shell powder prepared from matured shell is preferred to other similar materials like wood bark powder because of its uniformity in size and chemical composition. Shell charcoal prepared by burning shell in limited supply of air finds application in plywood and laminated boards as a phenolic extruder, and filler in synthetic resins, mosquito coils and agarbathi industry. Shell charcoal has extensive demand in manufacturing of activated carbon which is considered superior for gas absorption due to small-pore size and high mechanical strength.

❑❑❑

CHAPTER

24

ARECANUT

Botanical Name: *Areca catechu*

Chromosome No.: 2n = 32

Family: Arecacaae (Palmae)

Origin: Malaya, Phillipines, East Indies

Distribution: India, Bangaladesh, Sri Lanka, Indonesia, Malaysia.

Area	Production
World - 5.88 lakh ha	7.27 lakh tons
India- 3.00 lakh ha	4.78 lakh tons

In India it is cultivated in Maharashtra, Karnataka (50%), Kerala, Assam etc.

Uses

- Mastigatory
- Chewing along with betel leaves & tobacco
- Increase the production of saliva and gastric juices.
- For treatment of leucoderma, leprosy, cough, fits, anaemia, obesity etc.
- Being a source of tannin, it is used for dyeing clothes and tanning leather.
- Used for manufacturing plastic, hard boards etc.
- Leaf sheath is used for manufacturing ply boards, picture mounds, use and throw cups and plates.

Botany

Unbranched erect growing palm , height 60-70 ft, stem is smooth with scars of fallen leaves. Has 60-100 leaves partly fused and partly free. Inflorescence is a spadix with male and female flowers. Monoecious, fruit set 30%.

Cultivars: Mangala, Sumangala, Sreemangala.

Mangala- Semi tall, early bearing (3 years after planting) high nut set, good chewing quality.

Sumangala- Tall variety, medium sized oval shaped nuts, high percentage of chali recovery.

Sree mangala- compact bunches with bold and round nuts.

Variety	Country of introduction	Year of release	Ripe nut yield Kg/Palm	Chali(Dried kernel) Kg/palm
Mangala	China	1972	8.82	2.02
Sumangala	Indonesia	1985	12.9	3.3
Sree mangala	Singapore	1985	12.8	3.1

Other Varieties

Variety	State	Chali yield (Kg/palm)
Mohitnagar	West Bengal	3.7
Mettupalayam	Coimbatore	2.0
Thirthahalli	Karnataka	2.0
Sree vardhan	Maharashtra	2.0

Soil: Deep, well-drained soil.

Temp- 14°C to 36°C, Altitude- 1000m MSL

Propagation: Seeds

- Selection of mother palms: Early bearing, stabilized yield, regular bearing, high percentage of fruit set, 8-9 leaves/palm.
- Selection of nuts: Above 35gm, should float vertically when immersed in water.
- Primary nursery: Whole seed nuts are sown at 5 cm distance in vertical position with calyx and just covered. Germinates in 50-100 days.
- Secondary nursery: After 6 months in primary nursery planted at 30cm spacing. Nursery should be under shade.
- Seedlings of one to one and half years old are transplanted in the main field. Seedlings with more than five leaves, four nodes, 20cm height are selected. Planted in 90X90X90cm pits during May- June.

Spacing: 2.7 X 2.7m

Fertilizers: NPK 100: 40: 140kg/ha

2 splits May- June(1/3) Sep-Oct(2/3rd).

Intercrops: Banana, Betel vine, Tapioca, Pepper, Pineapple, Coconut, Jack fruit.

Mixed crop: Pepper, Cocoa, Banana, Lime, Betelvine.

Harvest: The climber climbs at one palm, harvest and then pull the nearest palm with a hook and swing to it to harvest the nut. One climber may harvest 100 palms by practice. The harvested branches are lowered to the ground using a rope.

Yield of ripe nuts: 700 to 1800kg/ha (Average)

Processing

Neetadakka- Method of keeping the nuts fresh conditions by steeping in water. Fresh fruits are washed with 100ppm chlorine water and blanched in 0.2% $Cacl_2$ solution. Kept immersed in a solution of 0.1% sodium benzoate and 0.2% Potassium meta bisulphite. In this method, fruits can be stored fresh for one year.

Kotapak or Chali: Only ripe fruits are harvested. Sundried for 40-45 days. Chali recovery 25%.

Grades of Chali: Moti, Srivardhan, Jamnagar, Jini.

Tender nuts: Harvesting green fruits at 6 months maturity. After dehusking, soft nuts are cut into pieces, boiled with dilute extract of previous boiling and drying. The same water is reused for 3 to 4batches. The extract is concentrated to make **"Thick kali"**.

Types (Based on cuts and shapes):

Api or Unde - Processed without cutting

Batlu- Cut transversly into two halves

Choor- Several longitudinal cuttings

Podi- Nuts are cut both longitudinally and transversly 3-4 times.

Scented supari: Prepared from chali and kalipak. Cut into bits and scented with essence of spices, synthetic flavor etc.,

Pest and disease: Spindle bug, root grub, white mite, fruit rot, foot rot (Anabe), inflorescence die back and yellow leaf disease.

❑❑❑

CHAPTER

25

CASHEW NUT

Botanical Name: *Anacardium occidentale*

Chromosome No.: 2n = 42

Family: Anacardiaceae

It is a dollar earning crop.

Origin: Brazil. Introduced to India by Portugese during 16th century.

India- Area is 9.23 lakh ha. Production is 6.13 lakh tons (Largest producer and processor of cashew)

Main production centres: India (150Mt), Brazil (75mT), Vietnam (71mT)

Exports: Kernels- 100 to 110mT, 50% of the world export is accounted by India. Exported to 60 countries.

Major importers: US, Netherland, Russia, UK.

India- AP, Maharashtra, Orissa, Karnataka, TN.

NRC for cashew – Puthur, Karnataka.

The **Cashew Export Promotion Council (CEPC)** of India was established in 1955.

- To promote export of cashew kernels, cashew nut shell liquid and allied products.
- To provide useful commercial information and to assist exporters.

Uses

1. **Kernel- A health food:** Cashew nut contains protein, fats and vitamins to a high degree and is equal to milk, eggs & meat. It is rich in amino acids which is absent in other nuts. Cashew kernel contains 47% fat- but 82% of

this has unsaturated fatty acids which is heart friendly and lowers cholesterol in blood. Cashew does not add obesity and help to control diabetes. Good appetizer, excellent nerve tonic besides exotic taste.

2. **Cashew nut shell liquid (CNSL):** The shell is 1/8 inch thickness, covering the kernel is a membrane. A honeycomb structure is inside the shell and contains a dark reddish viscous liquid – CNSL (Pericarp fluid of the cashew nut). CNSL contains two unsaturated phenolic compounds. Anacardic acid 90% and cardinal 10%. On roasting gives cardanol (Monohydric phenol)

CSNL has versatile industrial applications – friction lining, surface coatings, resins etc., It is heat resistant, water repellent, anti-corrosive. India exports CNSL and cardanol to USA, Japan, South Korea.

Cashew Apple

It is juicy, rich in carbohydrates & Vitamin C. It has astringent & acrid principles. Juice, Blended juice, beverages, juice concentrate, Candy, Jam, Pickle, Chutney, Vinegar, Cashew feni, Wine can be prepared.

Botany

Andromonoecious. It is a low spreading evergreen tree with a low spreading evergreen tree with a number of primary & secondary branches. Growth is cyclic.

Flushes emerge immediately after harvest

↓

After a month undergoes quiescent stage for 3 to 4 months

↓

Maturity

↓

Mature lateral shoots produce vegetative and reproductive flushes

↓

Panicle emerge on reproductive flushes

↓

Fruit set & harvest

Cultivars

- Ullal- Early cultivar. (Karnataka)
- Vengurla 1,2,,3,4,5,6,7- by Regional fruit Research Station, Vengurla, Maharashtra.
- VRI-1, 2,3, 4- RRS, Vridhachalajm, TN.
- Bapatla-1,2, 8- Cashew Research Station, Bapatla, AP.
- Madakathara- Kerala

Soil

'Gold mine' of wasteland. Altitude 700m, protects soil from erosion. Mostly grown on laterite, red soils, costal sands. Does not withstand water stagnation.

Climate

Tropical. Unable to tolerate frost or cold weather for a long period. Rain fall 50cm/ annum, not less than 20° C. Less in rain fall during flowering and dry period during nut maturity is ideal. Sun loving crop does not tolerate shade.

Propagation: Most of the old plantations are seedling progenies. Soft wood grafting gives 70-80% success.

Spacing: 8X8 m (175 plants)

Fertilizers: NPK – 500:125:125g/plant applied in 2 splits during May- June and Sep-Oct.

Intercropping

Andhra Pradesh	CN+ Casuarina+ Coconut
Kerala	CN+ Pineapple
Karnataka	CN+ Tamarind
W. Bengal	CN+ Bamboo

Pest and Disease

Tea mosquito bug- *Helopethis ant*oni

- Stem borer
- Die back or Pink disease
- Anthracnose

Harvest

Economic bearing 3rd year onwards. During Feb- May fully mature nuts are harvested. Nuts are picked up after they fall from tree. Apples are removed and sun dried for 2-3 days to reduce moisture from 25 to 9%.

Yield: Average- 2kg/ tree (3-5 years), 10kg/ tree (16-20 years)

Hybrids- 17Kg/tree (5 years)

Cashew Processing

Processing of cashew is defined as the recovery of edible meat portion- the kernel from raw nuts, by manual or mechanical means. In India, the processing is by manual means. It consists of:

1. Roasting 2. Shelling 3. Peeling 4. Grading 5. Packing

Moisture Conditioning or Humidifying

A slight under roasting or over roasting adversely affects the quality of the kernels. This is achieved by a moisture step preceding the roasting. The raw nuts are sprinkled with water and allowed to remain in moist condition for about 24-48 hours. This step is known as conditioning. The optimum moisture level at the end of roasting is reported to be 15-20%. Two important points to be taken care of during conditioning are;

- The water should not seep through the brown testa.
- The water should be free from iron contaminations.

Iron contamination in the water can interact with polyphenolic materials of testa and the resultant bluish black complex may give patches on white kernel.

1. Roasting: Roasting is designed to make shell brittle.

a. **Open Pan Roasting:** The earliest process was the pan roasting wherein the nuts are heated on a metal pan over an open fire. Due to the heat and slight charring the shells become brittle. The pan roasting is not followed in organized sectors of industries. The two important methods of processing now adopted are; i. Drum roasting and ii. Oil bath roasting

i. **Drum roasting:** The nuts are fed into a rotating hot drum, which ignites the shell portion of the nut. The drum maintains its temperature because of the oil oozing out of the nuts. The drum is kept in rotation by hand for about 2-4 minutes. The roasted nuts which are still burning are covered with wood ash to absorb the oil on the surface. The rate of shelling and the outturn of whole kernels are very high in this method. However, the main disadvantage is the loss of CSNL which has a very high export potential. In addition there will be considerable heat and acrid fumes in the vicinity of this operation.

ii. **Oil bath roasting:** In this method, the nuts are held in wire trays and are passed through a bath of cashew shell oil maintained at a temperature of 200-202oC for a period of three minutes whereby the shell oil is received from the shells to maximum possible extent. The vessel is embedded in brick work and heated by a furnace which use spent shell as fuel. During roasting, the shell gets heated and cell wall gets separated releasing oil into bath. As the level rises the oil is recovered by continuous overflow arrangement. The roasted nuts are then converted into a centrifuge. The residual oil adhering to the surface of nuts is removed by centrifuging. The roasted nuts are mixed with wood ash and sent for shelling. In Panruti (Tamilnadu) the conventional roasting is completely avoided. The raw nuts are exposed to the intense sun that is prevalent in that region. The well dried nuts are hand shelled. Here also the CSNL is completely recovered.

2. **Shelling:** After roasting, shelling is done by labour. Each nut is placed edgewise and cracked open with a light wooden mallet and the kernel extracted with or without wire prong. Care has to be taken that the inner kernel is intact and not broken into bits. After kernels are removed from the shells they have to be dried to reduce the moisture to loosen the adhering testa.
3. **Peeling:** Peeling is the removal of testa from the kernels. This is done with help of safety pin or small hand knife. Peeling is made easier when the kernels are subjected to a heat treatment for about 4 hrs in a drying chamber.
4. **Sweating:** After peeling, the kernels are spread out indoors on cement flooring so that they may absorb some moisture and become less brittle. This prevents the tendency to break easily during grading.
5. **Grading:** The next stage in the processing is the grading of kernels on the basis of specifications for exportable grades. There are 25 exportable grades of cashew kernels. The kernels are stored into wholes, splits and Broken primarily on the basis of visual characteristics. The wholes are again size-graded on the basis of the number of kernels per 1Ib. The entire grading operation is done manually. However for size-grading mechanical operation is also practiced.

Specification for cashew kernels

Grade designation	Number of kernels per lb	Grade designation	Number of kernels per lb
W 180	375 to 395	W 320	660 to 705
W 210	440 to 465	W 400	770 to 880
W 240	485 to 530	W 450	880 to 990
W 280	575 to 620	W 500	990 to 1100

General Characteristics: Cashew kernels shall have been obtained through shelling and peeling cashew nuts, shall have the characteristic shape, shall be white or pale ivory or light ash in colour, reasonably dry, and free from insect damage, damaged kernels and black or brown spots. They shall be completely free from rancid kernels. The kernels shall be completely free from testa.

Grade designations and their trade names:

Grade designation	Trade names		Trade names
SW	Scorched wholes	SB	Scorched butts
SSW or SW IA	Scorched wholes seconds	SS	Scorched splits
DW	or scorched wholes IA	SP	Scorched pieces
B	Desert wholes	SSP	Scorched small pieces
S	Butts	SPS	Scorched pieces second
LWP	Splits	DP	Desert pieces
SWP	Large white pieces	DSP	Desert small pieces
BB	Small white pieces	DB	Desert butts
	Baby bits Cont.	DS	Desert splits

6. **Packing:** Final operation is packing in 10 kg capacity tins, which are subsequently evacuated and filled with carbon dioxide. In some parts to overcome the possible over-drying a re-humidification step is introduced before packing. The practice of filling with an inert gas is mainly to combat infestation during transit. It may be pointed out that with high quality nut, free from infestation, storing with or without carbon dioxide makes very little difference particularly with reference to rancidity. The importance of inert gas appears to be more for circumventing a possible insect attack from an occasional insect egg entering the tin while packing.

   Nitrogen can also do the same function. However, carbon dioxide being a heavier gas is more convenient for handling. Contention that absorption of carbon dioxide makes the kernel tastier does not have much truth. In any case the processed kernels are rarely consumed without a subsequent heat processing in the form of roasting frying and/or baking.

By Products of Cashew

After the processing of the shell and other left outs are used making some other products. The major bye products of cashew processing are:

1. Cashew Nut Shell Liquid 2. Shell charcoal

Cashew Nut Shell Liquid (CNSL)

The pericarp of the nut consists of a coriacious epicarp, spongy mesocarp and stony endocarp. The kernel covered with testa membrane is contained in a shell

1/8 inch thick. The mesocarp consists of a honeycomb network of cells containing a viscous liquid called cashew nut shell liquid (CSNL), which provides a natural protection to the kernel against insects.

CSNL is a valuable raw material for a number of polymer based industries like paints and varnishes, resins, industrial and decorative laminates, brake linings and rubber compounding resins. CNSL is traditionally obtained as a byproduct during the isolation of kernel. The major constituent of shell oil is cardanol and anacardic acid of which cardanol is separately extracted and used in many industries. The shell oil was used as a preservative for boats and nets and to protect wood from termites. It is now largely exported and used in the manufacture of plastics, indelible inks, water proofing composition and other industrial products.

Extraction of CNSL

The extraction of CNSL involves various methods viz. hot oil bath, expellers, kiln method, solvent extraction etc, the most common method being hot oil bath. In this method the raw nuts are passed through a bath of CNSL itself by which the CNSL is extracted. This method extracts only 50% of liquid contained in nuts. Then through expellers about 90% of liquid can be extracted.

Cashew Shell Charcoal

The remains of shell after the extraction of CSNL is called shell charcoal. This is used as a fuel. The shell charcoal is used in processing of cashew for drying after shelling.

❑❑❑

CHAPTER

26

OIL PALM

Botanical name: *Elaeis guinensis*

Chromosome No.: 2n = 32

Family: Arecaceae

Palm Oil (or) Palmolein

It is the highest edible oil yielding crop. The fleshy mesocarp and the seed contain 45-55% oil. Therefore two types of oil *i.e*, palm oil and palm kernel oil are produced. Both are used for culinary and industrial purposes. Crude palm oil is the richest source of carotenoid. Oil palm produces the highest oil yield per unit area (2500- 4000 kg/ha) while it is 650-1500 kg/ha in coconut and 350-450 kg/ha in groundnut.

Origin: West Africa.

India- During 1961, commercial cultivation began in Kerala by importing 4000 trees from Nigeria.

World: Area - 12.6 million ha.

Production - 17.3 million ha.

Major producers : Malaysia and Indonesia-.

India: Area-8.00 lakh ha.

AP, Karnataka, Tamil Nadu.

Uses

- Manufacture of vanaspathy
- Does not increase blood cholesterol

- Anti- oxidant
- Rich source of beta- carotene, vitamin E
- Manufacture of soap, cosmetics, gums, candles, animal feed etc.,

Botany

Bears 50 leaves at the crown. Two leaves are produced per month. Monoecious with male and female inflorescence. A female inflorescence contains about 100 spikelets having 4000 flowers. In male inflorescence, each spikelet bears 600-1200 male flowers. Cross pollination occurs. *Elaeidobius kamerunicus* and *E. singularis* are the pollinating species which helps in high fruit set. *Thrips hawaiiensis* is also an important pollinating agent. Fruit is a drupe. The fruit ripens six months after pollination. Female fruiting bunches weighing 30-40 kg bears 2000 individual fruits. The fruitlets are black in colour when young turning to orange red when ripe. The palm produces one bunch per month and an average of 4 tons of crude oil palm is produced in one hector.

Cultivars

1. **Dura-** Thick shell (2-8mm), low mesocarp. Fresh Fruit Bunch to palm oil extraction ratio is 17%.
2. **Pisifera-** Shell- less with pea like kernels used for commercial planting.
3. **Tenera-** Hybrid between Dura & Pisifera. Shell is thin (0.5- 4mm). Fresh Fruit Bunch to oil extraction ratio is 20-26%. High mesocarp.

Soil

pH- 6.5 to 7.5. Heavy clay soils with poor drainage and sandy soils are not fit. Temperature influences inflorescence, flower initiation, sex differentiation, anthesis, floral abortion and growth. More male flowers are produced at low temperature. Best oil yield is obtained at 24-33°C.

Propagation

Seeds. Tenera hybrids are used. Well matured fruits are harvested. Fruits are separated from spikes by hand. After removing mesocarp and exocarp by knife, fruits are immersed in water for 10 days. Seeds are dried for two days. Seeds can be stored for 3-9 months at 27°C. The dried seeds are put in polybags and kept in germinator at 40°C. After 80 days seeds are removed and soaked in water for 5 days changing water daily and then dried for 2 hours. Then the seeds are kept in a cool place in poly bags. germination starts in 15-20 days. 90-95% germination is achieved.

Planting

It is advisable to plant well grown seedlings of 10-16 months age in the field. A well developed tenera seedling will have a height of 1-1.5 m from base and will have more than 13 functional leaves. Planting has to be done during the onset of rainy season. Pits of 60 x 60 x 60 cm^3 size are dug out maintaining both row to row and plant to plant distance of 10 m.

Irrigation

Continuous soil moisture encourages vigorous growth and increases yield of oil / palm. Water deficiency adversely affect flower initiation, sex differentiation and therefore, results in low sex ratio due to production of more male inflorescence.

Manures and Fertilizers

NPK: 1200: 600: 1200 g/ palm/year. The fertilizers are applied in two equal split doses during pre-monsoon (May-June) and post-monsoon (September-October). Magnesium deficiency can be corrected through the application of 500 g of magnesium sulphate per palm per year.

Boron deficiency is occasionally found in young palms in the field showing a reduction of leaf area in certain leaves producing incipient 'little leaf', advanced 'little leaf' with extreme reduction of leaf area and bunching and reduction in the number of leaflets and causing 'fish-bone' leaf. The 'fish-bone' leaves are abnormally stiff with leaflets reduced to projection. Leaf malformations including 'hook leaf' and corrugated leaflets are some other associated symptoms. Soil application of 50-200 g borax per palm, is given for correcting the malady.

Cover Cropping

Establishing cover crops in an oil palm plantation is advantageous. A mixture of three species of legumes viz., *Pueraria phaseoloides, Calapagonium muconoides* and *Centrosema pubescens* (2: 2: 1) is recommended. Seeds are mixed and sown in drills in between the rows of palms at the rate of 5 kg per hectare. Intercropping with cocoa and banana is quite successful. The crops like vegetables, flowers, cotton and tobacco are compatible with the plantation

Ablation

The inflorescences produced initially are very small and have low oil content. Removal of such inflorescence is called ablation. Removal of all inflorescences during the initial three years is found to improve vegetative growth of young palms and favours commencing of regular harvest after three and half years of planting. Ablation is done at monthly interval by pulling out the young inflorescence using gloves or with the help of devices such as narrow bladed chisels. Ablation also improves drought resistance capacity of young palms by improving shoot and root growth especially in low production areas where dry condition exists.

Pruning

Initial pruning has to be carried out just before assisted pollination begins, maintaining two fronds below the lowest developing bunch. The first pruning has to be carried out which is largely for very old fronds when the lowest part of the bunch is 30 cm above the ground. Further pruning should be done when the lowest developing bunch is 90 cm from the ground. The frond should be removed only at the time of harvest. When fronds are beyond the reach from the ground, fronds are removed with harvesting knives attached to bamboo poles. Two fronds should be left below the bunches maintaining a minimum of 37 fronds per palm.

Artificial Pollination

This is necessary to ensure fertilization of all female flowers in the bunches specially during rainy season. Pollens are collected from male inflorescence and are dried in sun for about five hours and stored in moisture free containers. It remains viable for over 30 days. The pollens are mixed with chalk powder (1 : 14) and are used for pollination. The pollen can also be used even without mixing with chalk powder by dusting them on receptive stigma.

Pests : Rhinoceros beetle, *Red palm weevil*

Diseases : Bud rot *(Ermima* sp.)

Harvesting and Yield

First harvest can be done 3.5 to 4 years after planting. Few ripe fruits are loose/ fall off indicates the bunch is ready for harvest. In Young palms, Chisel harvesting is followed. If palms become taller (from 10th year onwards), then harvesting is done by hook. Harvesting rounds should be made as frequent as possible to avoid over ripening of bunches. Harvesting rounds of 10-12 days are generally practiced. During rainy season harvesting rounds of 6-7 days are followed.

Yield: 25 tonnes of fresh fruit bunches (FFB) per hectare (Irrigated condition)
15 tonnes of fresh fruit bunches per hectare (Rainfed)

Processing

(i) Harvesting and handling of raw material.

(ii) Sterilization

(iii) Stripping

(iv) Digestion

(v) Oil extraction

(vi) Clarification

(vii) Oil purification and drying

(viii) Oil storage and transport

The functions of the above enumerated unit operation is mentioned below in brief:

(i) Handling of raw material: Oil palm fruit contains an extremely active enzyme called lipase which splits the oil into undesirable free fatty acids leading to quality problems during storage, processing and refining. This enzyme is activated by excessive handling and subsequent bruising while harvesting and transportation. During these processes care must be taken that:

a) There may be minimum damage to the fruits and the

b) Transportation of the harvested fruits may be done with minimum delay so as to minimize the damage due to lipase enzyme.

(ii) Sterilization: The purpose of this operation is to inactive the enzyme lipase and to loosen fruits to facilitate subsequent stripping. Also helps to soften the tissue for better digestion, to coagulate protein and to partially dehydrate the mesocarp.

(iii) Stripping: This process involves the separation of the fruits from the bunch stalks. Rotary drum stripper, which is rotated, due to centrifugal force the bunches fall back on to the bottom, scattering the fruits.

(iv) Digestion: The purpose of digestion is to disrupt the mesocarp and to break oil-bearing cells to facilitate oil release. This process also loosens the nut from the fibre.

(v) Oil extraction: Screw press is employed for oil extraction. The hot digested mass is fed to the press continuously and pressure is developed by compression. The oil water mixture is expelled through the perforations while the pressed cake along with nut is discharged through the other end of the screw.

(vi) Clarification: Crude oil from the pressing process contains about 50% water and non-oily solids, which must be removed. Separation of the oil from water and impurities is achieved at temperature of 90-95°C in a clarifier.

(vii) Oil purification and drying: Bulk of the water and other impurities are removed by a high speed centrifuge to bring down the moisture level. Further reduction of moisture is achieved by vacuum drying.

(viii) Storage and transport: Palm oil requires careful handling during storage and transportation, if quality deterioration is to be avoided. Such deterioration results in colour fixation, bleaching problems, off-flavours and loss of oil during refining.

❑❑❑

CHAPTER

27

PALMYRAH

Botanical name: *Borassus flabellifer*

Chromosome No.: 2n = 32

Family: Palmae

Government of Tamil Nadu recognized it as the State Tree of Tamil Nadu since 1978.

Origin and Distribution

The palmyrah is believed to be a native of Tropical Africa and grows extensively in the drier parts of India, Sri Lanka, Burma, Thailand, Vietnam, Malaysia and in some parts of Indonesia.

Botany

Palmyrah belongs to the genus Borassus. The distinguishing characters of the palm in this genus are their palmate, fan like leaves and dioecious character the male and the female flowers being borne on separate trees.

The palmyrah belongs to the species *Borassus flabellifer* and occurs throughout the dry areas of tropical Asia and Malaysia, being particularly abundant in India. Next to coconut, the palmyrah is believed to be the most abundant palms found in the world.

It is a perennial plant increasing by growth of new matter at the apex. It attains a height of 13 to 20 m and sometimes even 33 m. roots are slender and numerous. Trunk is arboreous, cylindrical and black in colour. In the early stages, only the underground portion of the stem increases in thickness and 15-20 years elapse before the aerial portion of the trunk elongates and develops into its characteristic cylindrical black stem. The outer portion of trunk is hard and composed of stiff longitudinal fibres. The central portion, the pith, is soft and starchy.

It bears a terminal crown of 30-40 large fan like leaves, of 1 to 2 m width. The leaf stalks are 60 to 120 cm long, strong and fibrous. The palmyrah leaf has 26-36 stalks (midribs). Inflorescence is axillary and spadiose. Spadix is branching and perianth is double whorled. Calyx consists of three sepals and corolla consists of three petals. It is a dioecious plant, the male flowers are embedded in fingers whereas the female flowers are sessile.

Botanically fruit is a drupe. A well mature palm on an average bears 50 fruits in bunches. Fruit is normally yellow to light orange colour with black bottom. It is spherical in shape with a hard green cap (bracts). Fruits are normally three seeded and the mesocarp is pulpy and fibrous and is protected by a thick fibrous skin (epicarp). The average weight is 1.25 kg and the pulp of which constitutes about 40 per cent.

Varieties

There is no recognized variety. They are generally classified as Black skin and Red skin fruited types. TNAU has released one improved variety viz SVPR-1 palmyrah palm from Srivilliputhur. It is a semi dwarf type with a high padaneer yield of 298 litres per palm in a tapping duration of 95 days. The padaneer of this variety has a high jaggery content (144 g per litre of padaneer) and a high brix content.

Propagation

The palm is propagated by self-sown seed. Most plantations have come up from self-sown and others by dibbling the seeds *in situ.*

Raising of Planting Material

The mother palm selected for seed nut collection should be free from pests and diseases, high yielders of padaneer and fruits, dwarf in stature, early and regular bearers, well matured fruit bunches indicated by the yellow tinge in the stylar and of the fruit may be harvested for seed collection. The selected seeds are stored in shade for 3 weeks. Shrunken, weight less and bored seeds may be rejected. The seeds may be directly sown *in situ* or sown in nursery to raise the seedlings. For direct sowing, 3 to 4 whole fruits are planted in pits half filled with sand or sand and soil mixture. The pit is then covered with dried leaves (mulch) to the same depth. Sowing may be taken up during rainy months. Seeds start germinating in three weeks.

Nursery Transplanting

If transplanting is to be undertaken, seeds can be sown in mound formed by heaping sand to form a bed of 1 m broad, 30 cm height and of convenient length or masonry nursery bed built with bricks of 2 m broad, 60 cm height and of convenient length. In both the cases, seeds are sown at 10 cm space and covered with 5 cm sand. About one year old seedling is lifted from the nursery and containerized in the polythene bags. After roots are formed the seedlings can be successfully transplanted in the main field. Shading the seedlings should be provided using coconut or palmyrah fronds.

Manuring

Farmers generally adopt sheep penning for their palmyrah plants since the soils are usually subfertile. Application of 10 kg of FYM /pit before planting the nuts (as basal dressing) is advantageous. The dosage may be increased biannually till reaching 60 kg FYM/pit/year.

Growth

The first frond appears in about five months. Production of fronds is slow to begin with. However the pace picks up as the age advances. The first fan shaped leaves appears only in the second year. It will be 6-7 years before the top of the trunk appears. This palm is a slow grower and grows to a height of 12-18 m and comes to flowering in about 13-15 years (irrigated condition) when it can be tapped for padaneer.

Pruning

Pruning of 30 per cent leaves at the rate of 10 per cent in a phased manner increased the yield of neera by about 28 per cent.

Tapping

The extraction of sap from the inflorescence is considered the most important use of this tree. Sap is extracted by a process referred to as tapping and there are different methods of tapping. These methods vary according to the sex of the palm and age of the inflorescence.

Tapping-male Palm

In the male palm, sap is extracted from the young (two weeks old) inflorescence by a method called **aripanai**. This method involves removing the sheath, which is covering the inflorescence and the inflorescence is allowed to dry for three days. The area around the inflorescence is also cleared by cutting the leaves. Unlike in other methods, no pressing or stroking of the inflorescence is carried out. A new surface is made by cutting and the pot is tied directly to the inflorescence. When the inflorescence becomes wet due to rain, the oozing of the sap stops immediately and inflorescence is discarded and the sap is obtained for one to one and half months by this method.

In the male palm, when the inflorescences are one month old, another method of extraction called **'Vallupanai'** is employed. Each male spike bearing sessile flowers is pretreated by pressing and stroking of the inflorescence.

Three to six spikes are brought together, wrapped with palmyrah leaves and fitted to a pot. The number of pots that are tied to an inflorescence varies from 2-4 depending on the number of spikes present.

Tapping - Female Palm

Sap is obtained from the female palm for a longer period (April to December) compared to the male palms (December to February). The young female inflorescence is tapped by a method called **'Thattupalai'**. Tappers soften the tissues by hitting the inflorescence's main axis with the iron rod and the fork is used to press at the regions from which the fruits develop. When the inflorescences are two or three months old, another method called **'Kaivetty'** is employed. The inflorescence at the stage is matured and bears fruits which are sliced as the tapping progresses.

The total yield of sap depends on the number of inflorescences per palm, the amount of work (pre-treatment) done on the inflorescence, genetic factor of the palm, environmental and edaphic factors such as wind, temperature, rainfall and soil. In the female palms, number of inflorescences per palm varies from 2 to 8 and in the male palms this number may vary from 4 to 10.

An inflorescence yields two liters of sap per day and that a large palm with several inflorescences in tapping will give about 20 liters per day, a single palm is estimated to produce 1,20,000 liters during the taping life.

Products from Palmyrah Palm

Edible Products

Neera

Neera tapped from the palmyrah tree is transparent, pleasant smelling and sweet. It is credited with medicinal values. It is used as a stimulant and antiphalegmatic and is also considered useful in inflammatory infections and dropsy. In many parts of India, Neera has been poor man's beverage from time immemorial. Neera was as effective as riboflavin injections in curing ariboflavinosis.

The sap contains 0.23 per cent protein, 12.6 per cent total sugar and small amounts of fat, mineral matter and vitamin C. The sap also contains most of the essential amino acids in the form of free amino acids. The sugar is mainly sucrose with trace amounts of glucose and fructose but during fermentation, sucrose is hydrolysed to glucose and fructose.

Toddy

Toddy is formed as a result of fermentation of the sugary sap with yeasts and bacteria which come into contact with the sap. This is an uncontrolled natural fermentation by a number of different strains of yeast and bacteria. The alcohol content in a fully fermented toddy is estimated to be about 5 per cent.

The major sugars present in partly fermented toddy are the sucrose, glucose and fructose but these are all gradually converted to ethyl alcohol during fermentation.

Palmyrah Arrack

The palmyrah toddy has been used to produce arrack. The approximate recovery of portable alcohol from toddy is 10 per cent.

Wine

Sterilized unfermented sap could be fermented with suitable strains of yeast to produce palmyrah wine. Sweet toddy, with pH adjusted to 6-7, sterilized and inoculated with good wine yeasts produce a very clear straw-coloured wine.

Jaggery

Jaggery is a solid mixture of reducing and non-reducing sugars prepared by concentration of palmyrah sap. Palmyrah jaggery with adequate amounts of calcium, the low values for sodium and high values for potassium propose that palmgur may be prescribed for dispensing conditions, like hypertension and oedema due to heart and liver disease. It can even be recommended for the treatment of disease with marked loss of potassium as in case of diabetic acidosis, post-operative convalescence and as a general diuretic.

Palm Sugar

Palm sugar can be substituted to cane sugar for all preparations. It has high dietetic values. It is considered to have healing properties.

Palm Cola

Palm cola is an aerated soft drink containing 11 per cent palm sugar.

Confectionary

Confectionary made out of sugar includes chocolates and ginger toffee.

Nungu

Like neera, the tender palmyrah fruits are called 'Nungu' and is relished as delicacy during summer by the people of Tamil Nadu. It is rich in carbohydrate, phosphorus, iron, niacin, vitamin C, Riboflavin etc.

Palmyrah Tubers

Three months old roots are consumed as food by the people in the regions where they are available. It is rich in carbohydrates. The tubers are cooked by direct fire or boiled in water with salt. Boiled tubers can be sliced, dried and stored. Such slices are powdered and various edible products are made.

Non Edible Products

Tender leaves: Used for making toys, flowers, garlands, hats and several fancy goods.

Mature leaves: They are used for making containers and packing materials.

Palmyrah fibre: The palm fibre is valued mainly for its high tensile strength leading itself to many industrial applications.

Timber products: Palmyrah timber is very valuable for the construction of house in palmyrah growing areas.

Yield

- 150 litres of padaneer / tree / year
- 24 kg jaggery/ tree / year
- Jaggery recovery/litre of padaneer: 180 – 250 g of jaggery

❑❑❑

CHAPTER 28

TEA

Botanical name: *Camellia sinensis* L. O. Kuntze.

Chromosome No.: 2n = 30

Family: Camelliaceae

Introduction

Origin – China / South East Asia

Area I - China (10.09 lakh ha)

II- India (5.78 lakh ha)

Production share I - India (28%)

II - China (22%)

Botany

Tea belongs to the genus *Camellia and* family *Camelliaceae*. Camellia though includes 82 species, only two species *viz.*, *C. assamica* and *C. sinensis* are the original species for tea. As they were highly crossable with each other, the present day tea seedlings (jat) are hybrids of these two species and are often referred as *C. assamica sub sp lasiocalyx*.

Tea is an evergreen shrub or tree. Leaves simple, alternate, serrated, flower bisexual, regular, sepals five, petals, five., stamens many, anthers two celled, ovary superior 2-4 locules, ovules 2-4 rarely solitary, axile, fruit is a capsule. In tea bush under plucking, a proportion of shoots will be actively growing, called as periodic shoots or flush shoots and the other shoots which remain temporarily dormant are called as banji shoots.

Periodic shoots develop from the axillary bud. It has definite pattern of growth. Aperiodic shoots develop from pruned bush frame. It grows in steady succession. The active growth of terminal bud prevents the growth of axillary buds, which occur in the axils of the leaves (cataphyllus), protecting the leaf primordia in the bud. The first scale leaf usually drops off with the scar seen at the base of every shoot. A malformed leaf growing like an ordinary leaf is called 'fish leaf' and has smooth margins and oval shape without prominent tip. Next, the normal leaves unfold in sequence from the growing bud.

4 weeks – sprouting of axillary bud

5-6 weeks – unfolding of scale leaf

6-7 weeks – unfolding of fish leaf

8-13 weeks – formation of pluckable shoot

Assam	China
Small Tree (1.5-1.8 m ht.)	Big shrub (90-150 cm ht.)
Few robust branches	Abundant, whippy shoots
Large, glossy leaves, 8-20 cm	Small, leathery erect leaves,
Long, 3.5-7.5 cm wide,	1.5-14 cm long and 1-2.5 cm
Light to medium green	Wide, Dark green colour
High yield,	Low yield
Cup quality-medium	Good quality
Less hardy, Susceptible to Drought and frost	Hardy and resistant to Drought and frost
Sparse flowering	Profuse flowering

Varieties

S.No.	Special characters	Clone
1	Wind Tolerance	UPASI-2 UPASI-10
2	Drought resistance	UPASI-9
3	Frost resistance	B-26
4	Smallest leaf	CH-1
5	Biggest Leaf	Betjan
6	Blister blight tolerance	TRI-2043,
7	High pubescence	TRI-2043
8	High anthocyanin pigmentation	TRI-2025
9	High tolerance to pH	TN-14-3
10	Poor fermenter	12/2
11	Mite tolerance	7/9
12	Scale insect tolerance	TN-14-3
13	High polyphenol content (53.7%)	Luxi white tea
14	High amino acid content (6.5%)	Anji white tea
15	High caffeine content (6.96%)	Wild tea at Yunnan

Improved Clones

S.No.	Clone	Important features
1	UPASI 1 (Ever green)	Hardy, Quality-Above average
2	UPASI 2 (Jayaram)	Hardy, Quality-Above average, tolerant to drought and wind
3	UPASI 3 (Sundaram)	Natural triploid quality clones and very high yielding
4	UPASI 6 (Brooklands)	Suited to mid and higher elevations
5	UPASI 8 (Golconda)	Suited to all elevations, high yielding
6	UPASI 9 (Arthrey)	Firly tolerant to drought and withstand slightly high pH, , high yielding
7	UPASI 10 (Pandian)	Hardy, Quality-Above average, tolerant to drought and wind
8	UPASI 14 (Singara)	Suited to higher elevations , High yield
9	UPASI 15 (Spring field)	Flushes throughout the year
10	UPASI 17 (Swarna)	Flourishing well at mid and high elevations
11	UPASI 24	Hardy
12	UPASI 25	High yielding
13	UPASI 16	High yielding
14	UPASI 27	Drought tolerant
15	UPASI 28 (UPASI 10 * TRI2025)	Biclonal, Good strength and high quality

Soil and Climate

Temperature required various from 16-32°C with well distributed annual rain fall of 125-150cm. Atmospheric humidity should be 80%. Tea requires well drained soil with high amount of organic matter and pH 4.5 to 5.5. The performance of tea is excellent at elevations ranging from 1000 - 2500 m. Tea is calcifuge crop requiring comparative low amounts of calcium and high amount of potassium and silicon.

Propagation

Tea is propagated by seeds and cuttings. Seed propagation is seldom practiced now.

Nursery

The nursery soil should be well drained and deep loam in nature with pH of 4.5 to 4.8. The soil and sand used in the preparation of rooting medium should be tested for pH and nematode infestation.

Pre-treatment of Rooting Medium

Treating with Aluminium sulphate can reduce soil pH. For this purpose the nursery soil is formed into beds of one metre width and about 8 cm height and of a convenient length. Then the beds are drenched with 2% solution of Aluminium sulphate applied at 10 litres/2.5 sq.m of area. Over this another layer of soil of 8 cm height is spread and again drenched with equal quantity of water twice. Then the soil is allowed to dry and the pH is checked before use in the nursery.

Preparation of Sleeves

Polythene sleeves of 150 or 200 gauge and 10 cm width and 30 - 45 cm length may be used. Drainage holes may be provided at the bottom. The lower 3/4 of the sleeves should be filled with 1:3 and soil mixture and the top 1/4 with 1:1 sand and soil mixture and staked in rows. Overhead shade is provided.

Selection of Mother Bush and its Treatment

Healthy and vigorously growing high yielding bushes should be selected. Apply to each selected bush with 40 g of young tea mixture + 60:90 NK mixture up to 5 years. The following mixture has to apply before taking the cuttings.

- 0.5 % $AlSO_4$ + 1 % $MgSO_4$ (before 3 weeks)
- 2 % Zn SO_4 (before 2 weeks)
- 1 % Urea (before 1 week)

Preparation of Cuttings

Cuttings are taken on April - May and August - September. Semi hard-wood cuttings are prepared with one leaf and an internode with a slanting cut at the bottom. Rooting was more in cuttings taken from two years old clonal plants than the old bush.

Planting of Cuttings

The sleeves are watered thoroughly and holes are made in the soil. The cuttings are inserted in the hole and the soil around is pressed firmly to avoid airspace followed by watering. Small polythene tents may be provided which maintain high humidity and regulate the temperature inside. Cuttings may take 10 - 12 weeks for rooting. After 90 days i.e. when all the cuttings have rooted, the polythene tent may be removed gradually over a period of 10 - 15 days.

Grafting

Even cleft grafting of single nodal cuttings is followed

Manuring of Nursery

After the tent is removed the cuttings are sorted and staked. 30 g of Nursery soluble mixture of the following composition dissolved in 10 litres of water may be applied over an area of 4 sq.m. This should be done fortnightly.

Hardening of the Cuttings

Hardening of 4 - 6 months old young cuttings should be done by removing shade gradually in stages over a period of 4 - 6 weeks starting from a few hours exposure to sun every day initially and extending the time of exposure gradually.

Methods of Planting

Single Hedge System: In this method, the spacing adopted is 1.20 x 0.75 m accommodating 10,800 plants/ha.

Double Hedge System: In this method, the spacing adopted is 1.35 x 0.75 x 0.75 m accommodating 13,200 plants/ha.

Season and planting - May - June or September - October

Irrigation

Subsoil irrigation may be given for young tea seedlings during summer months.

Manuring

Manuring should be done 2 months after planting. Phosphorous should be applied at 80 - 100 kg/ha as Rock phosphate once in a year by placement at 15 - 25 cm depth up to the first pruning and thereafter once in two years. N : K ratio 2 : 3 should be adapted for the first 3 years and a ratio 1 : 1 thereafter.

Year of application	Total weight kg/ha/annum		No. of applications
	N	K	
I year	180	270	5
II year	240	360	6
III year	300	450	6
IV year onwards	300	300	6

Application of fertilizers should be done before the onset of monsoon. Fertilizers should be broadcast around the drip circle avoiding contact with the collar.

AFTER CULTIVATION

Mulching

Covering the soil surface with any material is recommended. While mulching, care should be taken to keep the mulching material away from the collar region of the plants it reduces soil erosion, conserves soil moisture, suppresses weed growth and adds organic matter.

Weed Control

Perennial grasses can be controlled by spraying Glyphosate 1.75 lit + Kaoline 2 lit + 2 kg of wetting agent in 450 lit. of water followed by Gramoxone 500 ml in 200 lit of water to control dicot weeds.

Liming

In hill soils due to leaching of bases by rain and due to application of acid forming fertilizers, often soil pH is reduced. Hence periodical application of lime is essential to maintain optimum pH. Calcium carbonate or Dolomite is applied @ 1.5 t/ha prior to pruning (also dcpends on soil pH, rainfall, fertilizer usage and length of the pruning cycle).

Shade and its Management

Tea requires filtered shade and if it is exposed to direct sun, its growth is affected. Shade is hence essential and beneficial to tea as

1. It regulates the temperature.
2. It minimizes the effects of drought and radiation injury.
3. It increases the soil fertility (leaf litter adds about 8-10 tonnes of organic matter per ha/year).
4. It helps in recycling of nutrients.
5. It helps in getting even distribution of crop.
6. It serves as wind break.
7. It reduces the incidences of pests.
8. It generates additional income by way of timber and fuel.

The main drawbacks of shade are 1) increased incidence of blisterblight 2) competition with main crop for moisture and nutrients and 3) reduced response to applied fertilizers.

In South India, Silver oak (Grevillea robusta) is used as the permanent shade tree as it possesses the desirable characters of a good shade tree like

1. It must be an evergreen tree, easy to propagate having quick growing and deep rooted characters.
2. It provides filtered shade and withstands frequent lopping.
3. It tolerates wind and frost.
4. It does not have allelopathic effect.
5. It has commercial timber value also.

Six to nine months old seeds are ready for planting. Silver oak seedlings are initially planted along the tea rows at 6m x 6m spacing. As and when they grow, lower branches may be lifted periodically. When it attains 8 to 9m height, the tree is pollarded at a site having a girth of 10 to 15cm. Below the pollarded site, one branch in each direction may be left in 3 to 4 tiers and the excess ones are removed. Before every monsoon, the lateral branches, erect growing branches and shoot growth in the main stem are removed.

TRAINING YOUNG TEA

Centering

To induce more laterals, centering should be done 3 - 5 months after planting. Removing the growing point leaving 8 to 10 mature leaves from the bottom is done to induce secondaries.

Tipping

Tipping is done at a height of 35 cm from the second tipping at 60 cm from ground level.

When the secondaries reach more than 60 cm, they are tipped at 50-55 cm height by removing 3 to 4 leaves and bud to induce tertiaries. Therefore, plucking at mother leaf stage is continued for better frame development. It takes nearly 18 to 20 months from planting to reach regular plucking.

Pruning

Normally tea bushes establish themselves within a month of planting. In the course of one year or sometimes 18 months, they reach a height of 60-75 cm. After this stage, pruning may be taken up with the following objectives:

1. To minimize the height for purpose of easy plucking young buds, leaves and shoots.
2. To maintain the plant permanently in a vegetative phase.
3. To stimulate the production of young shoots
4. To keep the height of bush within bounds at easy and efficient plucking
5. To form an extensive frame as possible in two dimensions, on which in the third dimension, the bush will rapidly and continuously regenerate.
6. To renew the actively growing branches so that replacement of healthy wood and foliage keeps pace with ravages of death or damage.
7. To maintain sufficient volume of mature foliage to meet the physiological needs of the plant and
8. To promote rapid renewal of flush suitable for manufacture of quality tea.

Pruning is normally done 4 to 6 years interval depending upon the altitude of the garden, nature of the tea materials etc. The bushes marked for pruning should have adequate starch reserves in roots otherwise the sprouting following pruning will be less. This can be normally tested by the common Iodine test and if the starch reserve is less, bushes are allowed to rest for 2 to 3 months. The different types of pruning are as follows:-

S. No.	Type of pruning	Pruning height (cm)	Season	Remarks
1.	Rejuvenation pruning	20–China Jat30–Assam Jat	April-May	Done in old bushes affected with canker and wood rot to invigorate the new healthy branches. Not done regularly.
2.	Hard pruning	30-45	Apr - May	First formative pruning done to a young tea for proper spread of bushes
3.	Medium pruning	45-60	Aug-Sept.	Normal pruning wherever frames are healthy to stimulate new wood
4.	Light pruning	60-65	Aug-Sept.	Pruning depends on the previous history of the bush raising the height of medium pruning by an inch or less to manageable heights for plucking (less than 65 cm).
5.	Skiffing	65	Aug-Sept.	This is the lightest of all pruning methods. A removal of only the top 5 - 8 cm new growth is done so as to obtain a uniform level of pruning surface (more than 65 cm).

Pruning is done to maintain convenient height of bush and to remove dead and diseased branches.

$$\text{Area to be pruned every year} = \frac{\text{Total extent of the garden}}{\text{Pruning cycle}}$$

Pruning interval = (Elevation in feet / 1000) + 1

Pruning should be done in April - May or August - September

Immediately after the rejuvenation or hard pruning, the cut ends are smeared with a paste made of copper oxychloride and linseed oil (1:1).

Plucking

Plucking consists of harvesting 2 to 3 leaves and a bud. It is the most labour intensive operation in a tea industry and also decides the yield and quality of made tea. Normally, a pluckable shoot takes 60 to 90 days for harvesting since its sprouting from the axillary buds. When the shoot is plucked upto mother leaf, it is known as light plucking and if it is plucked below mother leaf, it is called hard plucking.

Plucking 'two leaves and a bud' is known as 'full' and anything less than this is known as 'fine'. Inclusion of more than two leaves is termed as 'coarse'. The younger the leaves, more is the tannin content and hence the superior quality tea. The plucking interval and plucking standard in relation to cropping is given below:-

Cropping pattern	Months	Plucking interval
High cropping or Rush cropping (60% of total crop)	April – June and October -December	7 – 10 days
Low cropping or lean cropping (40% of total crop)	July – September and January - March	12 – 15 days

It is essential to add one tier of active maintenance foliage to the bush every year. This is done by mother leaf plucking during January to March. During the rest of the period level plucking can be carried out.

Consequent to plucking, bush height increases every year in the order of 10 cm over tipping height in the first year, 7.5 cm, 7.5 cm, 5 cm and 5 cm over the previous year height in the second, third, fourth and fifth year respectively.

In some places, a scissor like mechanical shear harvester is employed to pluck during the high cropping period. It helps to manage the high crop and overcome labour scarcity.

Plant Protection

Major insects

1. Tea mosquito bug, Helopeltis theivora
2. Shot hole borer, Envallacea fornicatus
3. Red spider mite, Oligonychus coffeae
4. Scarlet mite, Brevipalpus californicus
5. Purple mite, Calacarus carinatus
6. Pink mite (or) Orange mite, Acaphylla theae
7. Yellow mite, Polyphagotarsonemus latus
8. Thrips, Scirtothrips bispinosus
9. Aphid, Toxoptera aurantii

Major diseases

1. **Blister blight** – *Exobasidium vexans*
2. **Red rust:** C*ophalearos mycoidea*
3. **Black root** : *Rosellina areuata*

Yield

Yield of made tea per hectare depends upon many factors such as elevation, clonal or seedling jats, management practices, severity of pruning, processing techniques etc., generally in tea industry, a field which yields upto 2000 kg of made tea/ha is considered as low yielding and 2000 to 3000 kg. as medium yielding and anything above 3000 kg as high yielding fields.

POST-HARVEST PRACTICES

Manufacturing of Tea

Basically, there are two types of processing *viz.,*

1. **Orthodox method** in which the rolling operation is done in a series of rollers. The rollers have rotary tables with battens, jacket for loading the leaf and a pressure cup.
2. **CTC method** (cutting, tearing and curling) which has a CTC machine, consisting of series of a pair of rollers mounted in such a way they rotate in opposite directions and the clearance between them is so adjusted to crush and tear the leaves. Irrespective of the method, manufacturing of tea involves the following steps,

a. **Withering:** The objective of withering is to reduce the moisture content of leaves by spreading them in troughs which receive artificial air from fan fitted on one end. At the end of withering, the leaves attain a flaccid condition for which it may take 12 to 18 hours depending upon the weather condition.

b. **Rolling:** This operation is carried on by a series of machines or in a single roller, during which the cells in the leaves are broken to liberate the sap containing the polyphenol oxidase, an enzyme, which in the presence of oxygen, oxidizes the polyphenols to produce theaflavins and thearubigens. These are responsible for colouring of the tea and are a prerequisite for next process *viz.,* fermentation. Rolling takes place for 30-40 minutes. Afterwards, the fine sifted rolled ones are sent for fermentation while the coarse ones are again sent for rolling.

c. **Fermentation:** Rolled tea materials are either spread in concrete floors or kept in aluminium trays. In the presence of high humidity and proper temperature, the properly fermented tea will take golden red colour. This step decides the quality *ie*. strength, colour and briskness of tea. Fermentation requires 1 hour or 2 hours depending upon the environmental conditions.

d. **Drying:** This step aims at stopping the fermentation process and slowly removing the moisture content without a burnt smell but preserving the inherent quality. This is achieved by passing the fermented tea in thin layers through conveyors into a drier in which the inlet temperature is maintained around 250-280°F and outlet temperature is a round 150-200°F. Proper drying takes 30-40 minutes.

e. **Grading:** Before grading, the dried tea is removed of the stalky fibres, which affect the quality, by passing through fibre separator machines. The bulk tea is passed through different sized meshes which aid in separation into different grades.

	Orthodox grades	Mesh size	CTC Grades	Mesh size
1.	Pekoe	>8 mesh sieve	Flowery Pekoe (FP)	>8 mesh
2.	Tippy Golden Orange Pekoe (TGOP)	8-12	Pekoe	8-10
3.	Broken Orange Pekoe (BOP)	12-16	BOP	10-12
4.	BOP-Fannings	16-18	Pekoe Fannings	12-16
5.	BOP-dust	18-24	BOP-Fannings	16-20
6.	Dust-I	25-30	Pekoe Dust (PD)	20-30
7.	Dust-II	Below 30	Red Dust (RD)	30-40
8.			Super Red Dust (SRD)	40-50
9.			Fine Dust (PD)	50-60
10.			Superfine Dust (SD)	Below 60

Green Tea

In green tea manufacture, the leaves after harvest are heated as quickly as possible which inactivates the oxidizing enzymes. Then they are cooled and rolled by further heating and rolling and eventually dried to 4% moisture.

Non-withered Tea

In this process, no withering is done. Firing becomes difficult and fermentation is incomplete leading to unfermented catechins and bitter taste. But it has good strength, colour and bright and uniform infusion.

Pickled Tea

Pickled tea is valued as a masticator in Burma. Wet leaves are tightly packed into mortar lined pits, compressed and made as silage.

Instant Tea

Water soluble extract of black tea is spray dried to powder form. It has high export potential.

Caffeine Free Tea

Caffeine from tea is removed by organic solvents and decaffeinated tea is marketed.

Oolong Tea

Oolong tea is semi or 50% fermented and contains polyphenol dimmers. It is produced only in China.

Silver tips Tea

Silver tips are generally produced by plucking only buds and drying them in natural sunlight.

Handmade Tea

Handmade tea is processed from shoots containing only one or two leaves and a bud. Plucked leaves were subjected to a hard wither of 40 to 50%.

Herbal tea

Herbal tea where green tea is mixed with medicinal plants, flavoured tea mixed with flavours of cardamom, bergamot mint etc. and scented tea mixed with jasmine flowers and organic tea are also produced.

CHAPTER

29

COFFEE

Botanical name: *Coffea sp.*

Chromosome No.: 2n = 22

Family: Rubiaceae

- One of the most agricultural export products.
- Second most important after petroleum products in the international trade.
- Exports and imports of coffee are controlled by International Coffee Organisation.
- According to Coffee Act, Coffee grown anywhere in India is the property of coffee board and thus the export and international trade of coffee is controlled by Coffee board.

Origin: Ethiopia and S.Africa.

Distribution: Even before 1200A.D it was consumed in Aden & Mecca. Commercial cultivation started in 1729. Since then Brazil is the major supplier ($2/3^{rd}$) of coffee to the world. Coffee was introduced in 1670 India by a pilgrim, Baba Budan who brought 7 seeds from Yemen and raised near Chickmangalum (Kar).

World: 10.55m ha, 7.7 mill.T

India: 3.2 lakh ha, 2.7mill.T

World- Brazil(2/3), Abyssinia, Columbia, Guatemala, S.India etc.

India: Karnataka (52% area), Kerala (28%), AP, Tamil Nadu

Species and cultivars (50-60 in the genus)

Four/ Five species contribute to commercial coffee

- *Coffea arabica* (2n=44) Tetraploid, superior, high altutudes.
- *Coffea robusta* (2n=22) Diploid, Inferior, Low altitude.
- *Coffea liberica* (Tree coffee):

12-15cm height, fruits 2.5cm dia. Thrives in lower altitudes Berries do not drop after ripening. Widely used for hybridization and a root stock for grafting *C.arabica*. generally planted in the borders of estate.

Distinguishing Feature

	C.arabica	*C.robusta*
Elevation	1000-1500m	500-1000m
Temperature	15-25°C, cool	20-30°C-hot, humid
RH	70-80%	80-90%
Annual Rainfall	1600-2500mm	1000-2000mm
Blossom showers	Mar-Apr	Feb-Mar
Backing showers	Apr-May	Mar-Apr
Shade	Medium-light	Light shade
Fruit maturity	8-9 months	10-11 months
Yield	2500-3000kg/ha	1250-1750kg/ha
Pollination	Self	Cross

Selection and hybridization been done by CCRI-Central Research Institute, Chikmaglur.

Varieties

Varieties	Parentage	Special characters
Selection 1 (S288)	Tetraploid hybrid from S26 (natural cross – *C. liberica x C. arabica*)	Resistance to leaf rust race 1 & 2, high yielding. Wider adaptability
Selection 3 (S795)	Cross between S.288 x Kent	Resistance to leaf rust race 1 and 2, 700 - 1200kg/ha. Bold fruits 75% A grade
Selection 5	Devamachy x S -881	Small, oblong fruits 900 – 1100kg/ha
Selection 6	S-274 (Robusta) x Kents	High A grade beans, 900 – 1000kg/ha
Selection 7	San Ramon –a short internode arabica	Dwarf in stature, segregates to tall by 30%
Selection 8	Pure line selection of Hibrido –de -Timor	Highest resistance to leaf rust, drooping branches
Selection 9	Sln 8x Tafarikela	Drought hardy, suitable to different coffee zones
Selection 10	Double cross hybrid, Cattura x(S795 XHDT)	Resistant to leaf rust
Selection 11	Progency of C.liberica X c.eugenoides	Field resistance to rust and drought hardiness

[Table Contd.

Contd. Table]

Varieties	Parentage	Special characters
Cauvery	Catimor x Hibrido de Timor	Plants are dwarf, suitable for high density plantingYield – 3000kg/haMore a grade beans with superior cup quality
Selection 12	Cross Caturra x HDT, followed by selfing	Precocious, suitable for close planting, resistant to leaf rust

Popular Arabica selection : sln 3, 5, 6, 8 , 9, cauvery.

Karnataka: Chickmaglur, Mudigere, S.coarg.

TN- Pulneys(s.5,6, 9), shervaroys (s. 795, s.9).

Botany

Perennial, evergreen. It has a prominent vertical growth from which sub lateral shoots develop. The vegetative growth of a year determines cropping in the next year.Therefore crop regulation should be done by pruning. Root system is shallow.

It is a short day plant. Floral initiation occurs during Sep- Dec. Flower buds are produced in the axils. Inflorescence is cymose. 4-5 inflorescence each with 1-4 flowers are found per axil (Arabica). Robusta has 5-6 infl/axil. Flower buds grow to a length of 7-8mm and remains quiescent until it is stimulated by blossom showers. Adverse weather high temperature leads to Star-flower- They are small, fleshy and green. Rain before and after flower opening affects pollination. Wind, gravity, bees are pollination agents.

Duration of flower to develop into fruit is 6-8 months in Arabica and 9-11 months in robusta. Fruit is a 'drupe' and contains two seeds. Sometimes abortion of one ovule leads to formation of single seeded fruit called 'pea berry' seed coat is a silver skin.

Bean Defects

1. Pea berry- single seeded
2. 3 or more seeds formed in a single fruit- Triangular seeds. Elepant beens – hollow and bit / defective.
3. Endosperm shows partial development – Floats/ Jollu

Nursery

Propagation through seeds. Seeds heve no dormancy. Seeds are sown in raised nursery beds in rows at 3cm apart. Season Dec-Jan. they germinate in 45 days.

At button/ topee stage, they are transplanted to secondary nursery/ poly bags after nipping off tap root.

Spacing

Arabica – Talls- 6' X 6', 7' X 7', 7' X 6'

Dwarf – 5' X 5'

C X R – 8' X 8', 9' X 9'

Robusta – 10' X 10' / 12' X 12'

Planting

- Pitz- $45cm^3$ during Mar-Apr
- 6-8 months old seedlings are planted during Aug- Sep
- 16-18 months old (secondary nursery) seedlings are planting during June.

Shade Trees

Shade belts in E.W. direction at 20' spacing X 40' bet rows. Permanent shade trees- 30-40' distance. *Grevellia robusta* (Silver oak), *Erythrina lithosperma, Ficus sp., Citrus sp.*

Roubusta is planted at wider spacing of 10-12' comes to yield only after 6-7 years. Inter planting can be done with var. Cauvery which bears in 3 years at 5' spacing. Additionally 3230 plants can be in corporate and later removed after spread of robusta.

Manuring

Species	Pre- Blossom March $N:P_2O_5:K_2O$	Post – blossom May $N:P_2O_5:K_2O$	Mid monsoon August $N:P_2O_5:K_2O$	Post- monsoon October $N:P_2O_5:K_2O$	Total
Arabica					
Young coffee 1st year after planting	15:10:15	15:10:15	—	15:10:15	45:30:45
2nd and 3rd year	20:10:20	20:10:20	—	20:15:20	60:45:60
4th year	30:20:30	20:20:20	—	30:20:30	80:60:80
Bearing coffee 5 years and above for less than one tonne/ha crop	40:30:40	40:30:40	—	40:30:40	140:90:120

[Table Contd.

Contd. Table]

Species	**Pre- Blossom March $N:P_2O_5:K_2O$**	**Post – blossom May $N:P_2O_5:K_2O$**	**Mid monsoon August $N:P_2O_5:K_2O$**	**Post- monsoon October $N:P_2O_5:K_2O$**	**Total**
For one tonne / ha and above Robusta	40:30:40	40:30:40	40:30:40	40:30:40	60:120:160
For less than one tonne/ ha crop	40:30:40	—	—	40:30:40	80:60:80
For 1 tonne /ha and above	40:30:40	40:30:40	—	40:30:40	120:90:120

After Cultivation

Weeding and mulching should be done as and when necessary. Digging is done to a depth of 30 cm towards the end of monsoon (October - November). The weeds and vegetative debris are completely turned under and buried in the soil while the stumps are removed. This is known as the cover digging. In slopy areas dig trenches on the contour 45 cm wide and 30 cm deep of any convenient length. Prune water shoots and disease affected shoots.

Training

Gives frame work and proper shape to the plant.

1. Single stem
2. Multiple stem are followed.

Single stem: High of the bush is restricted to a convenient length by topping. Topping is removing the vertically growing main stem two inches above the nodes at a particular height.

Arabica tall- 2.5 ft / Dwarf- 3-4.5 ft.

First toping Arabica (9-12 months), Robusta (18-24months), second tire in Arabica is topped at 4.5- 5 feet (Arabica).

Multiple stem: It is easy and cheaper.

Practiced in Africa/ Latin America. Once main stems are exhausted, new multiple stems are encouraged by stumping / bending. In India, multiple stem system is followed in replanted fields and HDP.

Pruning

To regulate cropping wood. Removal of old, unproductive, crisscross branches, diseased damaged branches etc. Remove new flush near 15cm around main stem during June-July and Aug-Sep.

Rejuvenation

Collar pruned/ stumped at 30cm above ground level after the receipt of summer showers in Mar- Apr. Cut surface should be treated with Bordeaux paste (1kg $CuSo_4$ + 1kg lime + 4.5 lit of water.)

Harvesting

Coffea Arabica- Comes to bearing in 3-4 years. Full production after 6-8 years. Harvesting period is 4 months for Arabica and 5 months for robusta from Nov-Feb.

Harvest

Harvest starts during November and harvesting extends up to February. Coffee fruits should be harvested as and when they become ripe. Coffee is just ripe when on gently squeezing the fruits the beans inside come out easily. Unripe fruits should be scrupulously sorted out before using the fruits for pulping. They may be dried separately as cherry.

Fly picking: Small scale picking of ripe berries during October to February

Main picking: Well formed and ripened berries are harvested during December. Bulks of the yields are obtained from this picking.

Stripping: Picking of all the berries left irrespective of ripening.

Cleanings: This is collection of fruits that have been dropped during harvesting.

Unripe fruits should be scrupulously sorted out before using the fruits for pulping. They may be dried separately as cherry.

Yield : Indian average is 946 kg/ha. / Arabica- 2500 to 3000 kg/ha / Robusta-1250 to 1750 kg/ha, Brazil- >3000 kg/ha.

Plant Protection

Major insects

1. White stem borer, Xylotrechus quadripes

2. Coffee berry borer, Hypothenemus hampei
3. Shot hole borer, Xylosandrus compactus
4. Mealy bugs, Planococcus spp
5. Green scale, Coccus viridis
6. Brown scale, Saissetia coffeae

Major diseases

1. **Rust** – *Hemileia vastatrix*
2. **Brown leaf spot** – *Cercospora coffeicola*
3. **Damping off / Collar rot** – *Rhizocotonia solani R.bataticola*
4. **Die back or Anthranose** – *Collectorichum coffeasum*
5. **Black rot or koleroga** – *Cortieium salminicolor*

Processing

Wet processing is done for Arabica and results in "Mild coffee". Dry processing is generally used for robusta and in Brazil 90% Arabica also.

Drying → Natural (Sun) –for 3-4 weaks upto 12% moisture

Drying → Artificial- below 60°C (riers – Static, rotary, horizontal, vertical)

Wet Processing

1. **Receiving**- Separate and remove floats, sand, stones, leaves, twigs etc.,
2. **Pulpling**- Removal of outer skin and pulp along with running water. Pulpers: Disc pulper, Drum pulper, roller pulper along etc.,
3. **Fermentation**- To hydrolyse mucilage and remove during washing. Some unwanted polyphenols will get dissolved and removed. Enzymes or alkaline solution is also added for repid removal of mucilage.
4. **Washing-** To remove mucilage. Horizantal washer, vertical washer are available.
5. **Drying-** Sun drying to get the find moisture content of 11%.
6. **Curing-** Hulling (Peeling)- Polishing- Grading- Sorting- Storage.

PROCESSING

Various Stages of Dry and Wet Processing

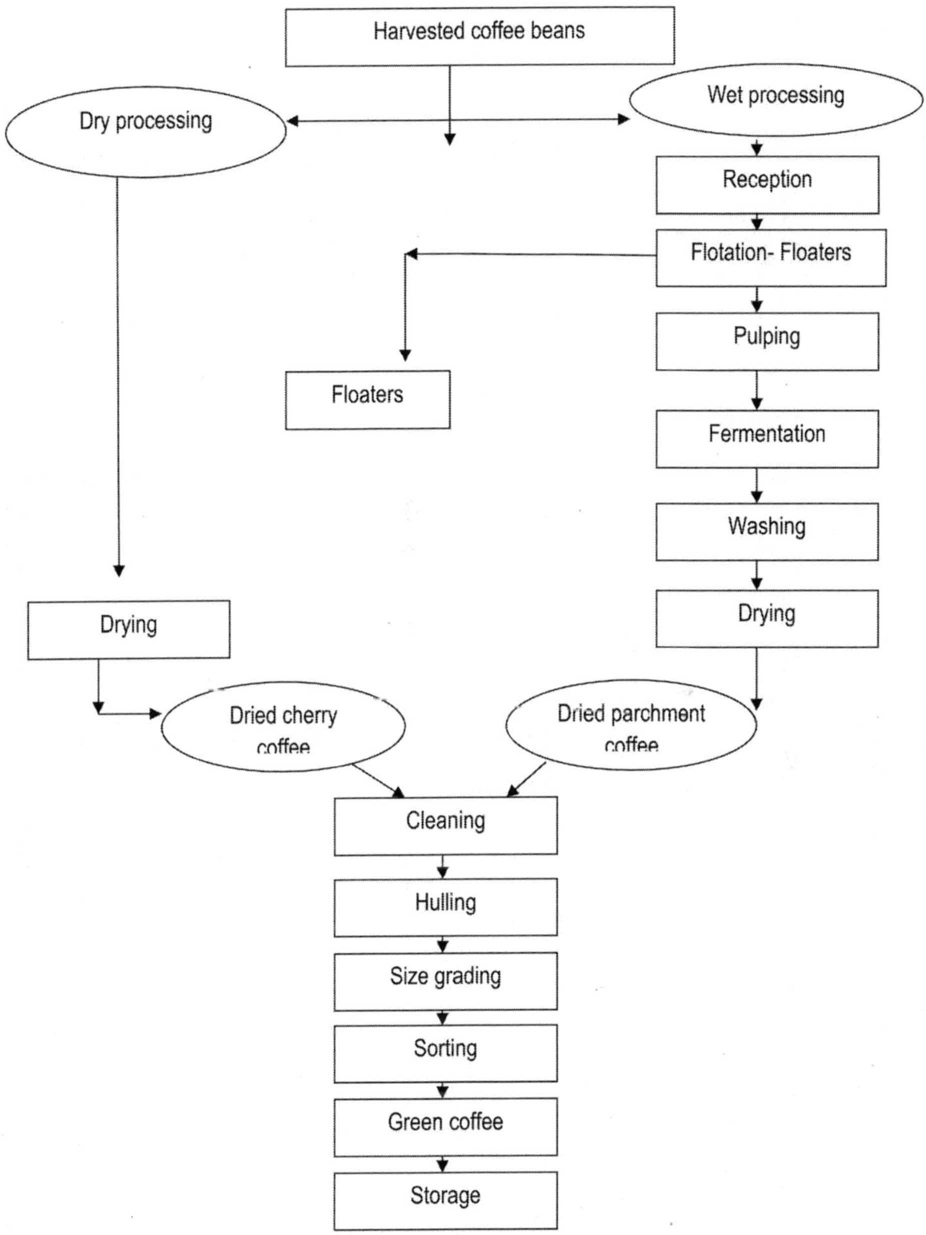

Fig. 36. Various stages of wet and dry processing

❑❑❑

CHAPTER

30

RUBBER

Botanical name: *Hevea brasiliensis*

Family: Euphorbiaceae

Chromosome No.: 2n = 36

Origin: South East Asia. In India during 1905 it was introduced into Kerala. Now mainly grown in tropical regions. Malaysia, Indonesia, India, Sri Lanka & other South East Asian countries.

Major areas: Indonesia (25 lakh ha)

Major producer: Thailand (25 lakh MT)

In India (4 lakh ha, 3 lakh MT) 90% of rubber is produced in Kerala and Kanyakumari. India is not self-sufficient. 60% of demand is met by import.

Major consumers: USA, China, Japan, India.

Composition: Yellow coloured opaque liquid. Natural rubber is obtained by latex, which has 30-40% rubber ash, with 60% water, protein, sugar, gum, tennin, pectin etc. These have stabilizing influence on rubber.

Uses:

- Auto tyres and tubes- 45%
- Cycle tyres and tubes- 13%
- Camel black, belts, hoses- 12%
- Foot wares- 11%
- Latex foams- 5%

Others: Shock absorbers, washers, gaskets, sports goods, contraceptives, household & hospital materials, paints, electronic, electrical, radio appliances.

Bitumin rubber powder- Road surfacing

Botany: Ever green tree, 18-30m height. Upright growth with crown like canopy. Wood is covered by bark, which on tapping yields latex. The cambium between bark and wood is responsible for increasing the girth of the tree as well as bark renewal. Moths, bees, flies are pollination agents.

Cultivars: Rubber Research Institute of India

Varieties	Yield	Special attributes
RRII- 105	Yield: 2460kg/ha/yr	Susceptible to drought.
RRIM- 600 (Rubber Research Institute of Malaysia)	1200 kg/ha	Tolerant to wind damage
RRIM- 703	1700 kg/ha	Tolerent to powdery mildew
GT-1, Indonesia	–	Thick bark

Soil: Hardy, pH 5.5 to 6.5, Sandy loam, forest loam & laterite soils ideal. Alkaline soil- Stunted growth.

Climate: Temp 20-28^{0} C, RF- 180 to 200cm/ann upto 600m, RH 80%, moderate wind & bright sunshine.

Propagation: Seed and bud grafting. Buds taken from high yielding trees grafted on seedlings.

Brown budding- Bud wood one year old, root stock 10 months old seedling.

Green budding- Green buds used on 2-8 months old seedling.

Crown budding- If clones are susceptible to disease/ wind, used for replacing clones.

Done at 2 to 2.5 m height.

Spacing: 5m X 6m

Population Buddings – 170 to 180 plants/ac

Seedlings- 180 to 210 plants/ ac

Initial population is high, selective thinning is done to get 134 plants/ac. One vigorous shoot is allowed to grow. Side shoots upto 3m should be removed.

Cover crops: to prevent soil erosion.

Calapagonium mucunoides, Mucuna bracteates

Shade : Albizzia & Gliricidia are used.

Fertilizer: NPK 40:90:90 g/plant/ year. Two splits May- June & Sep to Oct.

Pest and Disease

Pests: Scales, Mealy bug, termites.

Leaf fall: During south west monsoon due to fungal attack in leaf petioles. COC (Copper oxy chloride) spray at 0.25%.

Pink disease: Attacks stem- Exudation of latex, drip & run along stem. Mycelium forms "Cobweb" prolongs immaturing period. Spray- Bordeaux mixture.

Powdery mildew: Wettable sulfur.

Harvest

Tapping Height

Budded plant at 50cm girth, 125cm from bud union.

Seedlings at 50cm girth, 90cm height from ground.

Tapping

S/2 d/2	(half spiral, alternate day for 6 months and rested for 3 months)	100% Intensity
S /2 d/2	(half spiral, alternate day for 6 months and rested for 3 months)	67 intensity
S /2d/3	(half spiral, third day)	67 intensity
S/2 d/3	(half spiral, daily for one month and rested for next month)	100% intensity
S /1 d/4	Full spiral, fourth day	100% intensity
V /2 d/2	Half circumstances and cut alternate day for 12 months and rested for the next 4 months	75% intensity

Ethrel Treatment

Ethrel is recommended to increase latex yield of trees. It is applied at 5% a.i. concentration with a brush below the tapping cut to a width of 5 cm after light scraping of the outer bark. The first application may be done after a drought period preferably after a few pre-monsoon showers and subsequent applications may be done in September and November. However, continuous application of Ethrel is not recommended for periods of more than 3 years at a stretch.

Tapping Panel Dryness (Brown bast)

Syndrome characterized by prolonged dripping of latex with the gradual decline in volume yield, pre coagulation of latex and partial or complete drying of tapping area (10-25 per cent).

Control

Reduce tapping intensity or give a tapping rest for 3 to 12 months

Tapping Knife: "Miche Golledge"

Tapping cups:

i. Earthen ware- mostly used, cheap & heavy

ii. Glass, plastic(expensive, broken)- Rubber sticks to the cup.

iii. Cups are tied with wire and sprout is fixed to direct latex flow to cup.

Slaughter tapping: Before felling the tree for 2-3 years tapping intensity is increased without considering bark consumption and cambium wounding. Yd stimulants are also used.

Yield: Depends on volume of latex and % of rubber in latex.

Average yield- 850-2500 kg/ha/yr.

- Peak production – 15 years,
- Should be replaced in 30 years.
- Fresh latex is alkaline
- It becomes acidic rapidly due to bacterial action & coagulates.

Anti- coagulants- Ammonia 0.01%, sodium sulphate- 0.05%, formaldehyde- 0.02%.

❑❑❑

CHAPTER

31

COCOA

Botanical name: *Theobroma cacao*

Chromosome No.: 2n = 20

Family: Sterculiaceae

- Cocoa is a bevarage crop introduced in India in the early 1965's.
- Native of South America and widely cultivated in Ghana, Nigeria, Sierraleon, Cameroon, Brazil, Equador, West Indies and Malaysia.
- India it is cultivated in about 22,000 hectares as a mixed crop in Coconut and arecanut gardens.
- Kerala accounts for 79 % of the total area and 71% of the total production followed by Karnataka.
- Cocoa is highly cross pollinated

Botany

The leaves are alternate, entire, unlobed, 10–40 cm (4–16 in) long and 5–20 cm (2–8 in) broad. The flowers are produced in clusters directly on the trunk and older branches; they are small, 1–2 cm (1/2–1 in) diameter, with pink calyx. The fruit, called a cacao pod, is ovoid, 15–30 cm (6–12 in) long and 8–10 cm (3–4 in) wide, ripening yellow to orange, and weighs about 500 g (1 lb) when ripe. The pod contains 20 to 60 seeds, usually called "beans", embedded in a white pulp. Each seed contains a significant amount of fat (40–50% as cocoa butter). Their most noted active constituent is theobromine, a compound similar to caffeine. A shade-grown cocoa tree can produce fruit for 75 to 100 years or more.

Varieties

There are three broad types of cocoa - Forastero and Crillo, as well as Trinitario, a hybrid of the two. Within these types there are several varieties.

Forastero

Producing the greater part of all cocoa grown; Forastero is hardy and vigorous, producing beans with the strongest flavour. The Forastero variety most widely grown in West Africa and Brazil is Amelondaro. It has a smooth yellow pod and pale purple beans.

Crillo

With its mild or weak chocolate flavour, Crillo is grown in Indonesia, Central and South America. Crillo trees are not as hardy and produce softer red pods, containing 20-30 white, ivory or very pale purple beans.

Trinitario

Plants are not found in the wild as they are cultivated hybrids of the other two types. Trinitario cocoa trees are grown mainly in the Caribbean, but also in Cameroon and Papua New Guinea. The mostly hard pods contain 30 or more beans of variable colour, though white beans are rare.

Only Forastero types are known to perform well under Indian conditions. Breeding work initiated at the Kerala Agricultural University since 1979 has resulted in the release of seven improved clones of Forastero type. These are CCRP 1, CCRP 2, CCRP 3, CCRP 4, CCRP 5, CCRP 6 and CCRP 7.

During 2002, three hybrids viz., CCRP 8, CCRP 9 and CCRP 10 were released. These have mean yields of 90, 105 and 79 pods per tree per year. These are also tolerant to vascular streak dieback disease.

Climate and Soil

- The tree cannot withstand high winds, drought or sudden fall in temperature.
- The crop requires well-distributed rainfall.
- The minimum requirement of rainfall is about 100-150 cm per annum.
- Situations where the temperature falls below 10°C or rises above 38°C are unfavourable although minor deviations from the above limit can be adjusted by shade and irrigation.
- Cocoa is grown at altitude up to 900 m above MSL
- The best soil for cocoa is forest soil rich in humus. Shallow soils should be avoided. Cocoa is grown on soils with a wide range of PH from 6-7.5

Selection Criteria for Planting Materials

Cocoa can be propagated by seed and vegetative means.

Seed Propagation

It is desirable to collect seeds from biclonal or polyclonal seed gardens involving superior self-incompatible parents to ensure genetic superiority of planting materials. Polyclonal and biclonal seed gardens have been established at CCRP farm of the Kerala Agricultural University, Vellanikkara and Kidu farm of CPCRI and seeds and seedlings are being supplied to growers. If seeds cannot be procured from such seed gardens, mother plants for collection of seeds may be selected based on the following criteria:

a. Trees of Forastero type having medium or large pods of not less than 350 g weight or 400 cc volume, green in colour when immature, having smooth or shallow furrows on the surface without prominent constriction at the neck should be selected. Yield of pods should be not less than 100 per year.
b. Husk thickness of pods to be not more than 1 cm.
c. Pod value (number of pods to give 1 kg wet beans) to be not more than 12.
d. Number of beans per pod to be not less than 35.
e. Bean dry weight to be not less than 1 g.

Seeds lose viability within a week of harvest of pods. Seeds are to be sown immediately after extraction from the pods. Viability of the beans can be extended for some more days if freshly extracted seeds are stored in moist charcoal and packed in polybags. Other alternative is extracting beans, removing the testa and packing in polythene bags.

Selection of Planting Materials

When seedlings are used for planting, select only vigorous and healthy seedlings produced from polyclonal seed garden or selected mother plants as described earlier.

When budded plants are used, select two or more clones for planting as the use of a single clone can lead to poor production due to the existence of self-incompatibility in cocoa.

SEEDS AND CULTIVATION

Selection of Site

Cocoa is usually planted under coconut and arecanut plantations in India. Shade levels under coconut canopy are highly variable depending mainly on the spacing of coconut, extent of canopy development and age of palms. It is estimated that light infiltration through coconut canopy ranges from about 30 to 80 per cent depending upon these factors. Based on this, the general recommendation is as follows:

1. If a choice is possible, a coconut plantation that will let in more light through the canopy may be chosen for raising cocoa.\
2. If the light infiltration is over 50 per cent, it may be beneficial to provide additional shade using temporary shade plants like banana.

Preparation of Land

The seedlings / budded clones are usually planted in the interspaces of coconut / arecanut. Give a spacing of 3 to 4.5 m. The crop is best grown with 50 per cent light intensity in the early stages. In the early life of the plants, planting of quick growing plants like banana and tapioca can provide temporary shade.

Time of Sowing

Though the seeds will germinate at any time of the year, seeds may preferably be sown by December-January, so that 4-6 month old seedlings become available for planting by May-June.

Method of Sowing

Seeds are to be sown with hilum-end down or to be sown flat. Sowing is to be as shallow as to just cover the seeds with soil. Removal of pulp may enhance the speed of germination, but the extent of additional advantage is only marginal. Seeds start germination in about a week and germination may continue for another one week. Percentage of germination may be around 90.

Cocoa nursery is to be located in a heavily shaded area, which allows only 25-50 per cent sunlight. Regular watering is necessary to keep the soil moist.

Seedlings are transplanted after 4-6 months. Only vigorous seedlings are to be used and based on height and stem girth, 25% poor seedlings may be rejected. When seedlings are grown under heavy shade, hardening for 10 days by exposing to higher illumination may be necessary before transplanting.

Vegetative Propagation

In view of the high variability exhibited by seedling progenies, vegetative propagation is preferred for large scale planting. Though vegetative propagation of cocoa by budding, rooting of cuttings and grafting are feasible, the widely accepted method in India is budding.

Scions for budding are to be collected from high yielding, disease resistant elite plants. Shoots having brown bark and just hardened leaves are selected as bud wood. Scions are preferably procured by cutting off lamina of all the leaves of the selected scion shoot to a distance of about 30 cm from the tip. After 10 days when the petioles have fallen off, these scion shoots are cut and used for budding immediately. Bud wood can be stored by dipping in benzyl chloride followed by washing in water and then sealing the cut ends using molten wax. Bud wood is then wrapped in moist cotton wool and in turn in wet tissue paper or blotting paper and packed in boxes with wet packing material. The packet is then covered using polythene sheets. Storage life of the bud wood can be extended up to 10 days by this method. As far as possible, bud wood is to be collected from chupons as those produced from fans may develop into bushy plants with spreading habit. Rootstock, six to twelve months old may be selected in such a way that scion and rootstock are of the same thickness. Different successful methods include T, inverted T, patch, and modified Forkert methods. Patch budding is adopted in the Kerala Agricultural University.

Time and Method of Planting

Cocoa is planted as a pure, mixed crop or intercrop. When planted as a pure crop, Dadap (*Erythnina lithosperma*) is planted at 3x 3m spacing to provide shade. Dadap needs pruning every year. For more permanent shade, *Albizzia stipulate* can be planted adopting 9x9 or 12x12m spacings. This requires 4 to 6 years to develop proper canopy to provide sufficient shade.

Cocoa is planted as an intercrop in coconut and arecanut gardens. In coconut, depending upon the spacing adopted, one or two rows of cocoa can be planted in between two rows of coconut i.e., two rows where the spacing is more than 8 m and one row otherwise, the plant distance for cocoa being 2.7 to 3 m. When two-row system is adopted, the seedlings may be planted in zigzag or triangular manner.

In arecanut where the normal spacing is 2.7 m, cocoa is planted at the centre of four areca palms along alternate rows of interspaces only. Pits of 50 x 50 x 50 cm are dug, allowed to weather for one month and refilled with topsoil and 15-20 kg of compost or farm yard manure to ground level. The planting hole should be sufficient to hold the soil ball of the polybag. Tear off the polybags carefully, place the soil ball with the seedlings in the planting hole with minimum disturbance and press the soil around firmly. Planting should coincide with the onset of monsoon, but in places where irrigation is resorted to, flexibility in the time of planting is possible.

Shaping of Clonal Plants Derived from Fan Shoots

Budded plants from fan shoots have diffuse branching system and bushy growth habit. This type of growth causes difficulties in carrying out cultural operations and harvesting. If a better shape of the plant is desired, appropriate formation pruning may be necessary. This involves identification of a chupon arising from a fan shoot, allowing it to grow and removing the original, lower fan-like shoots in stages. This, however, has to be done slowly as an early drastic pruning will inhibit growth.

Pruning and Training

Cocoa grows in a series of storeys, the chupon or vertical growth of the seedling terminating at the jorquette from where four to five fan branches develop. Further vertical growth is continued through a side chupon that arises from a point just below the jorquette which again jorquettes after growing to some height. Left for it, the plant will grow to a height of 8-10 m repeating this process of jorquetting and chupon formation 3-5 times. When cocoa is grown as an intercrop in coconut and arecanut plantation, it is desirable to restrict the growth to one tier formed at a convenient height preferably above the head level of the workers. When jorquetting takes place at lower levels this can be raised by nipping off all the fan branches and allowing one chupon to develop and grow further to jorquette at the desired height. After this is achieved, further vertical growth is arrested through periodical removal of chupons.

The intensity of pruning is to be decided by the nature of growth of individual trees, shade intensity, growth of the companion crops etc. In the early stages, pruning is done to give a particular shape to the tree. After the establishment of the trees in the garden, prune them to the extent of retaining only the required number of leaves (20-30 leaves per developing pod). Removal of secondary branches from the centre should be restricted only to those trees growing in excess shade.

Top Working

This technique is useful to rejuvenate old and unproductive cocoa plants and also to convert genetically poor yielders to high yielders. This consists of snapping back the desired trees below the jorquette after cutting half way. The snapped canopy continues to have contact with the trunk. A number of chupons would arise below the point of snapping and this is triggered by the breakage of apical dominance and continued connection with the snapped canopy. Patch budding as described earlier may be done on three to four vigorous and healthy shoots using scions from high yielding, disease resistant clones and the remaining chupons are removed. The polythene tape is removed three weeks after budding and the stock portion above the bud union is snapped back. The snapped portion is removed after two hardened leaves develop from the bud. When sufficient shoots are hardened, canopy of the mother tree can be completely removed. Because of the presence of an established root system and the trunk with reserve food, the top worked trees grow much faster and give prolific yield one year after the operation. Though top working can be done in all seasons, it is preferable to do it in rain-free period in irrigated gardens. For rainfed situations, it may preferably be done after the receipt of pre-monsoon showers.

Top worked trees start yielding heavily from the second year onwards. About 50 per cent improved yield is obtained in the second year and about 100 per cent improved yield in the third year. Loss of crop for one year during the operation is compensated by bumper crop in the coming years. The main stem will continue to belong to the older plant and fruits borne on this area belong to the poor yielder. Better yields are however obtained from the fan branches of the high yielding clone used for top working.

Irrigation

Cocoa grows well as a rainfed crop under conditions of well-distributed rainfall and irrigation is not necessary. If sufficient moisture is not present in the soil due to prolonged drought or failure of rains, irrigation is to be given once in five days. Irrigation, however, helps in better growth of plants and precocity in bearing.

Manuring

Apply $N:P_2O_5:K_2O$ in two equal split doses in April-May and September-October, @ 100:40: 140 g / tree / year. $N:P_2O_5:K_2O$ may be applied @ 200:80:280 g / plant / year, in trees yielding more than 50 fruits per year. Dolomite @ 100 g / plant / year may be applied to plants from the third year onwards.

Under irrigated conditions, the yearly dose may be split into four and applied during April-May, September-October, December and February-March.

Apply 1/3 of adult dose during the first year of planting, 2/3 during second year and full dose from the third year onwards.

Apply fertilizers in circular basins with a radius of 25 cm during the first year. Gradually increase the radius of the basin to 120 cm by the third year. Apply fertilizers in the entire area of 1.5 m radius around the tree followed by forking in.

Plants showing zinc deficiency symptoms (narrowing of leaves, sickle leaf formation, green vein banding, and chlorosis in the interveinal areas) should be sprayed with 0.5 to 1.5% $ZnSO_4$ three times a year.

Weed Management

During the first three or four years after planting, it is essential to keep the field free from weeds. Maintenance and regulation of shade should be carried out promptly. During the establishment phase of the crop particularly in summer, provide mulching with materials like chopped banana sheath, coconut husk, cocoa husk etc. to conserve moisture in conditions of direct insolation. A mature cocoa plantation should form a proper canopy, which will be dense enough to prevent weed growth. Operations such as pruning and regulation of shade should be attended to in time.

Pest Management

- Stem borer (*Zeuzera coffeae*)
- Striped squirrel (*Funambulus sp.*)
- Rats (*Rattus rattus*)
- Myllocerus weevils (*Myllocerus viridanus*)
- Mealy bugs (*Planococcus citri and Rastrococcus sp.*)
- Aphids (*Toxoptera aurantii*)
- Cockchafer beetle (*Popillia sp. and Leucopholis sp.*)

Disease Management

- Seedling blight (*Phytophthora palmivora*)
- Phytophthora pod rot / black pod (*Phytophthora palmivora, P. citrophthora, P. capsici*) Lasiodiplodia pod rot / charcoal pod rot (*Lasiodiplodia theobromae*) Colletotrichum pod rot (*Colletotrichum gloeosporioides*)
- Phytophthora canker (*Phytophthora palmivora*)
- Pink disease (*Corticium salmonicolor*)
- Vascular streak dieback (VSD) (*Oncobasidium theobromae*)
- White thread blight (*Marasmius scandens*)

Harvesting

It takes about 170 days for a cocoa pod to develop from formation to maturity. During the period from 70-140 days after pollination, the size of pods and their fat and sugar content increase rapidly. Ripening takes about 25 days, during which, the pods change colour depending on the variety. Pods remain suitable for harvesting for fairly long time after they have ripened. Hence, it is possible to have harvest of sufficient number of pods at a time by either delaying the harvest of early-ripened pods or harvest of pods, which are fully ripened. Harvesting should be done at regular intervals rather than daily, once in 7-10 days. Avoid over-ripening of pods. The discards at the harvest can be left in the garden either in the open during summer or in pits at different sites in the rainy season, or they can be incorporated in the compost. Pod husks from the fermentary can also be used similarly as a good source of organic manure.

Pods are removed by cutting with a sickle-sharp knife, without damaging the cushion from which it is developed. After 2-3 days, they are split by banging them against some hard objects. Opening the pods with a knife damages the beans. During the period between harvesting and splitting, pre-fermentation activity inside the pod is hastened, which improves later fermentation. Beans from the split pods are scraped out with fingers. Portions of placenta and broken, germinated, caked, parthenocarpic and undeveloped beans are removed. On an average, 10-12 pods give 1 kg of wet beans and 3 kg of wet beans (from 30-36 pods) give 1 kg of fermented and dried beans.

Yield: 50-70 pods/tree/year

PROCESSING

Fermentation

During fermentation, the pulp or mucilage covering the fresh beans is removed and characteristic chocolate flavour is imparted to the final produce. The process is simple but must be carried out properly in order to get beans of good quality. Heat is produced by keeping the fresh beans compactly and this heat must be conserved so that chemical changes inside the bean can be completed. The four methods of fermentation usually employed involve the use of baskets, heaps, boxes and trays for filling up the wet beans.

Tray Method

The best method suitable for small quantities of beans is the tray fermentation. Wooden trays, 10 cm deep with slatted / split cane bottoms are divided into a number of sections by means of wooden partitions that fit into appropriate grooves at required distances. The capacity of the tray can be adjusted depending upon the availability of beans by keeping the wooden plank in the appropriate grooves.

A convenient tray can be 25 cm wide and 60 cm long. Wet beans are filled in the tray and levelled. About 10 kg of wet beans may be required to load one tray fully.

A single tray of beans will not ferment properly and at least four or five trays are needed for successful fermentation. The trays are stacked one over the other in such a manner that the cocoa filled portions are in a single row one above the other. The top tray is covered with plantain leaves. After 24 hours, a close fitting sack is put to cover the stack to keep the beans warm. Mixing or stirring of beans is not necessary and fermentation gets completed in 4 to 5 days, whereas 6 to 7 days are required for other methods of fermentation.

Basket Method

In this method, bean lots ranging from 2-6 kg can be fermented successfully. Mini baskets may be made of bamboo matting, closely woven and should have a diameter of 20 cm and height of 15 cm for a capacity of 2 kg. For slightly larger lots, proportionately deeper baskets may be used (e.g., for 6 kg, the depth may be about 40 cm). The baskets are lined with one or two layers of torn banana leaves to facilitate drainage of sweating. Wet beans are then filled, compacted and covered with banana leaves. The baskets are placed on a raised platform to allow the flow of drippings. After 24 hours, it is covered with gunny-sack and applied weights (bricks). The beans are to be taken out and stirred well 48 hours and 96 hours after the initial setting. Fermentation will be completed in six days and the beans can be taken for drying on the seventh day.

A number of factors influence the duration of fermentation. Weather changes and season are important through their influence on temperature and atmospheric moisture. Ripening also affects fermentation. Beans from unripe pods cannot be fermented. Beans of Criollo ferment more quickly than those of Forastero. During the early stages of fermentation, heat is produced by the action of anaerobic microorganisms. The beans are killed by the combined effect of heat and acetic acid and the cocoa aroma and flavour potential are developed.

Judging the End Point of Fermentation

Well-fermented beans will be plumpy and filled with a reddish brown exudate. The testa becomes loosened from the cotyledons. When cut open, the cotyledons will have a bleached appearance in the centre with a brownish ring in the periphery. When above 50% of beans in a lot show the above signs, it can be considered as properly fermented.

Drying

On completion of fermentation, beans are dried either in the sun or by artificial means. Sun drying can be done in thin layers 2-3 cm deep and stirring from time to time. Under normal sunny weather, drying can be completed in four to five days. While drying in mechanical driers, care must be taken to avoid exposure of the beans to smoke, fumes etc. The most common method of determining bean dryness is to take a sample and compress this in the palm of the hand and listen for the characteristic sound, which is associated with correctly dried cocoa. The more scientific method is to use a moisture meter.

Storage

The dried beans with moisture content of 6-8% may be packed in polythene bags or polythene lined gunny bags. Some special conditions have to be provided in storage in order to maintain the quality of the cured beans. Properly dried beans can be kept in 200-300 gauge polythene covers if only small quantities are involved or in polythene lined gunny bags in the case of larger stocks. Beans should be cleaned of flat, broken and other defective beans before storing. The store should be sufficiently ventilated and the bags should be kept on a wooden platform with air space of about 15-20 cm below the wooden planks set over the floor. The humidity should not exceed 80% so as to prevent mould development and pest incidence in the beans. Before storing cocoa, the store can be made clean and insect free by application of pesticides well in advance, but pesticides should neither be applied nor be kept with the beans inside the store. As cocoa beans can absorb and retain permanently any odour from its surroundings, other food-stuffs should not be kept with cocoa. So also, smoke or kerosene fumes should be prevented from entry.

CHAPTER

32

BETELVINE

Botanical Name: *Piper betel*

Family: Piperaceae

Chromosome No.: 2n = 26

Origin: Betelvine is a perennial dioeciously creeper native of Malaysia.

Importance

Betelvine leaves have a strong pungent aromatic flovour and are widely used as masticatory. Mature leaves are used for chewing with smeared hydrated lime plus catachu, arecanut, clove, cardamom, clove, funnel etc. Betel chewing is considered as a good and cheap source of distary calcium. Betel oil has several medicinal used it used for brain leaves carmination, etc. Leaves; has religious importance.

Composition

Moisture 85.4%, protein 3.1%, fat 0.8%, carbohydrate 6.1%, fibre 2.3%, minerals 2.3%, reducing sugars 0.38%, to 1.46% and all vitamins and Iodine also. It also contains almost all amino acids. The aromatic value is determined by the nature of oil content in leaf. Oil content varies from 0.7 to 2.6% the oil consist of phenols and terpens.

Varieties

Karpurakodi, Kallarkodi, Revesi, Karpuri, SGM 1, SGM (BV) - 2, Vellaikodi, Pachaikodi, Sirugamani 1, Anthiyur kodi, Kanyur kodi and Bangla type are under cultivation

Climate

Betelvine thrives best under humid tropical condition with cool shade. It flourishes in areas with a rainfall of 225 to 475 cm.

Soil

Well drained fertile clay loams are suitable. It does not tolerate saline and alkaline conditions.

Seeds and Sowing

The vines are propagated by terminal stem cutting or setts about 30 - 45 cm long. Setts obtained from the top portions of the vines are easy to root and hence best for planting. On an average 1,00,000 setts are required for planting one hectare. Setts with vigorous apical buds and nodal adventitious roots are selected and planted at the base of the live supports, which are to be planted 4 to 5 months earlier.

Row spacing	Vines/hectare	
	Single vine	Double vine
20 cm	50,000	1,00,000
30 cm (1 ft)	30,000	60,000
45 cm (11/2 ft)	22,500	45,000

Season

November - December and January – February are optimum for cultivation.

Preparation of Field

The field is prepared to a fine tilth and beds of 2 m wide are formed to a convenient length. Provide drainage trenches of 0.5 m width by 0.5 m depth in between two adjoining beds. Plant the seeds of the live supports *i.e.* Agathi (*Sesbania grandiflora*) in long rows. About 750 banana suckers are planted at the edges of the beds, which are used, for tying the vines on the live support and for packing the betel leaf. When the Agathi plants reach 4 m height, they are topped off for maintaining the height. The crop is planted in two rows in beds of 180 cm width on Agathi plants with a spacing of 45 cm between plants in the row.

Irrigation

Irrigate the field immediately after planting and afterwards once in a week.

After Cultivation

Training of the live standards

Before the establishment of vines, the side branches of Agathi trees upto a height of 2 m are removed for early creeping of the vines.

Training of the Vines

Training is done by fixing the vine at intervals of 15 to 20 cm along the standards loosely with the help of banana fibre. Training is done at every 15 - 20 days interval depending upon the growth of vines.

Lowering of Vines

Under normal cultivation, the vines grow to height of 3 m in one year period. When they reach this height their vigour to produce normal size leaf are reduced and they need rejuvenation by lowering during March - April. After the vine is lowered, the tillers spring up from the nodes at the bends of the coiled vines at the ground level and produce many primary vines. Irrigation should be given after each lowering.

Manuring

Apply 150 kg N/ha/year through Neem cake (75 kg N) and Urea (75 kg N) and 100 kg P2O5 through Super phosphate and 30 kg Muriate of potash in three split doses first at 15 days after lifting the vines and second and third dose at 40 - 45 days intervals. Apply on beds shade dried neem leaf or *Calotropis* leaves at 2 t/ha and cover it with mud (2 t in 2 split doses).

Time of application	N	P	K
		(kg/ha)	
Basal dressing	37.5	100	50
Top dressing @ 3 split doses	112.5	0	0

PLANT PROTECTION

Pests

Scale Insects

Select scale-free seed vines. Spray Chlorpyriphos 20 EC 2 ml/lit when one or two scales are noticed on the basal portion of the stem/leaves. Direct the spray solution to the basal portion of the vines. Spray NSKE 5 % or Malathion 50 EC 1 ml/lit.

Mites

Mites can be controlled by spraying Wettable sulphur 50 WP @ 1 g/lit or Dicofol 18.5 EC 0.5 ml/lit.

Sooty Mould (Aphids)

To control aphids spray Chlorpyriphos at 2 ml/lit on Agathi leaves. Clip off excess Agathi leaves.

Mealy Bugs

Mealy bugs can be controlled by spraying Chlorpyriphos 20 EC at 2 ml/lit or Dimethoate 30 EC 2ml/lit. Concentrate the spray towards the collar region.

Nematode

Application of Neem cake at 1 t/ha or shade dried *Calotropis* leaves @ 2.5 t/ha can be applied to soil for controlling the nematode populations.

Soil application of Bacillus subtilis (BbV 57) or Pseudomonas fluorescens @ 10 g / vine for the control of root knot nematode and quick wilt of betel vine

Diseases

Phytophthora Wilt

Integrated method for the management of *Phytophthora* wilt.

- Select well matured (more than 1 year old) seed vines free from pest and diseases.
- Soak the seed vines for about 30 mins in Streptocyline 500 ppm or Bordeaux mixture 0.5 %.

- Apply 150 kg N/ha/year through Neem cake (75 kg N) and Urea (75 kg N) and 100 kg P_2O_5 through Super phosphate and 30 kg Muriate of potash in 3 split doses first at 15 days after lifting the vines and second and third dose at 40 - 45 days intervals. Apply on beds, shade dried neem leaf or *Calotropis* leaves at 2 t/ha and cover it with mud (2 t in 2 split doses).
- Drench Bordeaux mixture 0.25% in basins formed around the vine at monthly intervals starting from October – January, three times soil drench and six times spray from June - July.
- During winter season avoid frequent irrigation.
- Application of *Trichoderma viride* @ 5 g/vine.

Bacterial Leaf Spot, Blight and Bacterial Stem Rot

Spray Streptocycline @ 400 ppm + Bordeaux mixture @ 0.25% at the time of first disease symptoms appear. Continue spraying at 20 days intervals. Always spray the chemical after plucking the leaves.

Anthracnose

Spray 0.2 % Ziram or 0.5% Bordeaux mixture after plucking the leaves after the first appearance of the symptom. The variety Karpoori is susceptible to the disease.

Powdery Mildew

Powdery mildew can be controlled by spraying 0.2% wettable sulphur after plucking the leaves.

Harvest

Harvesting of leaves starts from 6 months to 18 months after planting depending of soil, varieties, each vine is picked thrice or four times in a year.

Yield

About 75 to 100 lakh leaves/ha/year can be obtained.

❑❑❑

REFERENCES

Anbu , S,. S.Parthiban, J. Rajangam and T.Thangaraj. (2001). Induction of off-season flowering in mango (Mangifera indica L.) using paclobutrazol. South Indian Horticulture. Vol.49. Special issue. Pp.384-385.

Anbu , S,. S.Parthiban, J. Suresh and T.Thangaraj. (2001).High density system in mango (Mangifera indica L.). South Indian Horticulture. Vol.49. Special issue. Pp.384-385.

Anonymous. 2008. Area, production and productivity of horticultural crops. Published by Directorate of Hort. & Plantation Crops., Government of Tamil Nadu, Chennai. pp. 82 - 83.

Anonymous. 2013. National Horticulture Board.

Anonymous.2004.Mango. Crop production techniques for horticultural crops. Tamil Nadu Agricultural university.p.1-3.

Auxcilia, J and S.Sathiyamoorthy. (2001). Human hair developed amino acid as an organic substitute for nitrogen in papaya. South Indian Horticulture. Vol.49. Special issue. Pp.214-215.

Bose , T.K and S.K.Mitra. (1990). Fruits: Tropical and Subtropical, Naya Prokash, 206 Bidhan Sarani, Calcutta-6.

Chada, K.L. (2001). Hand book of Horticulture. ICAR, New Delhi-12

Chattopadhyay, T.K. (2008). A text book on pomology (Tropical Fruits- Vol-II). Kalyani Publishers. New Delhi.

Chezhiyan, N., P. Balasubramani, C. Vijulan harris and M. Ananthan. 1999. Effect of inorganic and biofertilizers on growth and yield of hill banana var. Virupakshi. *South Indian Hort.,*47 (1-6) 161.

Dinesh, M.R and I.S. Yadav. (1998). "Surya": A promising papaya hybrid. Indian Horticulture, 43(3):p.21.

Hayes, W.B. (1975). Fruit growing in India (III Edn.). Kitabistan, Allahabad.

Iyer, C.P.A., Y. Selvaraj, A.M.D.S. Subramanian and N.G. Divakar. 1976. Characteristic of mango fruits as influenced by exposure to sunlight during growing season. Prog. Hort., 8: 57 – 62. .

Jeyakumar , P., N.Kumar and K.Soorianathasundaram. (2001). Fertigation studies in papaya (Carica papaya L.). South Indian Horticulture. Vol.49. Special issue. Pp.71-75.

Kumar, M. 2012. Exploring the environmental, physiological, biochemical and molecular characterization of off-season flowering phenomena in mango (*Mangifera indica* L.) Ph.D. Thesis submitted to Horticultural College and Research Institute, Periyakulam.

Kumar, N and L.Nalina. (2001). Research on high density planting of banana in Tamil Nadu- A resume. South Indian Horticulture. Vol.49. Special issue. Pp.1-5.

Kumar, N. 2010. Introduction to Horticulture (Seventh edition). Oxford and IBH Publishing Co. Pvt. New Delhi.

Mahalakshmi, M., N. Kumar, P. Jayakumar and K. Soorianathasundram. 2001. Fertigation studies in banana under normal system of planting. *South Indian Hort.*, 49: 80-85.

Naik, K. C. 1949. South Indian fruits and their culture. P. Varadachary and Co., Madras.

Palaniappan, S.P. 1985. Cropping systems in the tropics – Principles and Management, Wiley Eastern Ltd., New Delhi, p. 215.

Parthiban, S., K. Balakrishnan, C. Rajamanickam and M. Selvarajan. 2009. Flowering behavior in mango. National seminar on production, post-harvest technology and marketing of mango. (Eds. Balasubramaniyam*et al*), Horticultural college and Research Institute, Periyakulam, India., pp . 105-108.

Rajan, S. 2009. Off season mango production: present status and future scope. In: National level training cum seminar on off-season mango production. (Eds. Richard Kennedy *et al.*), Horticultural Research Station, Pechiparai, India pp 1-4.

Rajan, S., Ramesh kumar and P.K.Pathak. (2003). Ambika – A promising mango hybrid: Spectrum. ICAR News. January-March. Pp.16-17.

S.Parthiban, J. Suresh , S. Anbu and T. Thangaraj. (2001). Effect of planting densities on yield of sapota (Manilkara achras Mill.). South Indian Horticulture. Vol.49. Special issue. Pp.25-26.

Shanmugavelu , K.G. (1987). Production Technology of Fruit Crops. SBA publications.